God said,

"Let there be lights
in the vault of heaven
to divide the day from night,
and let them indicate festivals,
days and years.
Let them be lights
in the vault of heaven
to shine on the earth."
And so it was.
God made the two great lights:
the greater light to govern the day,
the smaller light to govern the night,
and the stars.
God set them in the vault of heaven to shine on the earth,
to govern the day and the night
and to divide light from darkness.
God saw that it was good.
Evening and morning came: the fourth day.

THE FOURTH DAY

What the Bible and the Heavens Are Telling Us about the Creation

HOWARD J. VAN TILL

WILLIAM B. EERDMANS PUBLISHING COMPANY
GRAND RAPIDS, MICHIGAN

255 Jefferson Ave. S.E., Grand Rapids, Mich. 49503

Printed in the United States of America

Reprinted, January 1989

Library of Congress Cataloging-in-Publication Data

Van Till, Howard J., 1938–
The fourth day.

1. Creation. 2. Astronomy—Religious aspects—
Christianity. 3. Physics—Religious aspects—
Christianity. 4. Bible and evolution. I. Title.
BT695.V36 1986 231.7′65 85-29400
ISBN 0-8028-0178-1

Cover: The Great Nebula in Orion, located at a distance of 1500 light-years from us within the disk of the Milky Way galaxy. This giant cloud of glowing gas, approximately 20 light-years in extent, is populated with numerous newborn stars and star-forming regions.

Frontispiece: M 51, the Whirlpool galaxy. This spiral galaxy, accompanied by a smaller satellite galaxy, is about one-third the size of the Milky Way galaxy and is located at a distance of 12 million light-years from us.

Contents

To the memory of my parents
Clarence and Sena Van Till,
who placed a high value on
significant questions and meaningful answers

Preface

The biblical doctrine of creation is an essential component of the Christian faith. To know God as Redeemer, we must first know him as Creator. But what does it mean to say that God is our Creator? And what does it mean to say that the entire cosmos—from atoms to galaxies—is God's Creation? Must these questions be answered by giving a historical chronicle of specific creative acts? Or could they be answered by providing literary pictures of the eternal, covenantal Creator/Creation relationship?

What does the Bible tell us about the Creator and the Creation? What kinds of questions about the material world can appropriately be addressed to the Bible? In what ways do the cultural and historical contexts in which the Bible was written affect the form or the content of its answers to these questions? On what kinds of questions is the Bible silent? What does it mean to take the Bible seriously when it speaks about the material world? In sum, what does the cosmos look like when viewed through the spectacles of scriptural exegesis?

These are some of the theological and exegetical questions that have long been of interest to me. Born into a Christian home, I was given a thorough, integrated training in the Christian faith at home, in the church, in Christian elementary and secondary schools (in Ripon, California), and in college (Calvin College, at which I am now a faculty member). On the foundation of that training, I have continued to build my understanding of the Bible's answers to these questions. In Part I of this book I present what I have come—tentatively—to believe are the answers; naturally, I make no claim that they are complete or final.

There is another category of questions that have also aroused my curiosity for a long time. What is the physical nature of the cosmos? When we look through telescopes, or microscopes, or spectroscopes, what do we see? What physical properties does the material world display? What kind of behavior is exhibited by matter and material systems? What has happened in the past? What sequence of events and processes has preceded the present state of affairs in the cosmos? What is the scale of cosmic chronology? Is cosmic history evolutionary in character? What does it mean to take the material world seriously? In sum, what does the cosmos look like when viewed through the lens of scientific investigation?

As a scientist trained in physics and astronomy, I have spent the past two decades participating in the quest to find answers to such questions and evaluating the answers provided by the scientific community of which I am a part. In Part II of this book I present an overview of the scientific enterprise and provide samples of the results of research in contemporary astronomy, giving special emphasis to the physical properties, behavior, and temporal development of stars.

The Christian community, of which I am a member, rightly wishes to view the cosmos through the spectacles of scriptural exegesis—to see the cosmos as Creation. The scientific community, of which I am also a member, rightly wishes to view the cosmos through the lens of scientific investigation—to see the cosmos as a coherently functioning material system. These two ways of looking at the cosmos yield two distinctly different views. How valid or credible is each of these views? In what way do they differ from one another? In what way are they related to one another? Are they contradictory views, or do they complement one another? How may we as Christians take seriously both scriptural exegesis and scientific investigation? Is it possible to integrate these two views into a coherent, unified perspective on the cosmos? These questions concerning the relationship of the two views form the subject matter of Part III.

It is my contention that neither the scriptural nor the scientific view of the cosmos is complete in itself, despite the fact that each view contributes an essential perspective on the complete reality. Through the spectacles of scriptural exegesis, we Christians see the cosmos as Creation; we see where it stands in relationship to God the Creator, who is its Originator, Preserver, Governor, and Provider. Through the lens of scientific investigation, natural scientists are able to observe the internal affairs of the material world—its coherent properties, its lawful behavior, and its authentic history. Both views are integral parts of what I call the "creationomic perspective," the view of the cosmos that is gained when natural science is placed in the framework of the biblical doctrine of creation.

This book is addressed to those who want to take both the Bible and the Creation seriously, to those who, like myself, are vitally concerned first to get clear and accurate views of the cosmos through scriptural exegesis and scientific investigation and then to form a unified, coherent perspective that incorporates both views. Together we will seek a coherent perspective that is both biblically sound and intellectually honest. May God bless us in this effort. And may the result of our efforts bring praise to the Creator, whose glory the heavens are telling.

Acknowledgments

I am indebted to several persons and institutions for the aid and encouragement that they have given to me in the course of this writing project. I am particularly grateful to Calvin College for providing me with a sabbatical leave during the spring semester of 1982 to begin the research, reading, and reflection required to weave a unified perspective from many strands of thought, and for granting me a Calvin Research Fellowship so that I could use the summer and fall of 1983 to write the first draft of this book. I would also like to thank the National Endowment for the Humanities for supporting my participation in a Summer Seminar (1982) on "The Exact Sciences in Antiquity and the Middle Ages," led by Professor Asger Aaboe at Yale University.

In the development of Part I of this book, dealing with biblical interpretation, I benefited greatly from conversations with my colleagues, Clarence Vos, John Stek, and Raymond Van Leeuwen. I would also like to thank Professor Meredith Kline, of the Gordon-Conwell Theological Seminary, for his written suggestions regarding the first five chapters. While it was not possible for me to incorporate all of the suggestions contributed by these biblical scholars, I am grateful for their many valuable insights and for their personal encouragement. Let it be clearly understood, however, that I bear full responsibility for the final form and content of this section.

Thanks also to Dr. Paul A. Vanden Bout, director of the National Radio Astronomy Observatory, for his reading of most of Part II, thereby helping to ensure that the scientific discussion found in these chapters faithfully represents contemporary astronomy and astrophysics.

For a superb job of typing the entire manuscript and numerous revisions, I want to thank Jan Woudenberg, the mainstay of our science division office. And for his thorough editorial assistance and thoughtful encouragement, I express my deep appreciation to Tim Straayer of Eerdmans Publishing Company.

Finally, a special word of gratitude to my wife, Betty, my children, my students, my colleagues, and other friends who have encouraged me to think that this work may be helpful to the Christian community, and who have challenged me to speak as candidly and honestly as possible. To use a familiar line, Thanks, I needed that!

PART I

The Biblical View

The nearly full moon, photographed by the author at the Calvin College Observatory.

"When you raise your eyes to heaven, when you see the sun, the moon, the stars, all the array of heaven, do not be tempted to worship them and serve them."

—Deuteronomy 4 : 19

CHAPTER ONE

Taking the Bible Seriously

As I suggest in the preface, this work is addressed to readers who want to take the Bible seriously. My purpose in making that remark is not only to establish the assumptions I am making about the knowledge and attitude of my readers but also to indicate my own attitude and goal. I, too, want to take the Bible seriously in matters of God's work as both Redeemer and Creator. But what does it mean to take the Bible seriously in these matters? This is, clearly, a crucial question—the topic of much lively discussion and sometimes the subject of heated (but not always illuminating) debate within the Christian church. It is a question that deserves the best efforts of our theologians, who in turn deserve the support, encouragement, and prayers of all Christians.

Trained primarily in the natural sciences (more specifically, physics and astronomy), I make no claim to be a professional theologian or biblical interpreter. Yet each of us who reads and studies the Bible necessarily does so with some particular view of its nature and with some working principles for its interpretation. As I prepared to explore the Bible for its teachings concerning the celestial bodies, the sun, moon, and stars, I found it essential to begin by considering carefully these matters of the nature and interpretation of the Bible. The material in this chapter is my own attempt to formulate a working approach to scriptural interpretation.

What does it mean to me to take the Bible seriously? To

state it as directly as I can, it means to respect the Bible for what it is and to respond to it appropriately.

The Bible has been characterized in a multitude of diverse ways. We call it "holy Scripture" and the "Word of God." We may refer to it as "special revelation" or characterize it as "inspired" or "infallible." We see it described as "salvation history" or as a "written witness to God's redemptive acts" or as our "infallible guide for faith and life." This diversity of expressions illustrates both that Scripture is perceived in many different ways and that we use different descriptions or characterizations of Scripture to emphasize its many different aspects. Each of the descriptions given above may be considered accurate or true in an appropriate context, but because of the richness of the nature and character of the Bible, no one of them by itself exhausts Scripture's attributes. (Similarly, if I would say "My wife, Betty, is a talented organist," it would indeed be a true statement, but it would fall far short of fully describing her.)

Clearly, taking the Bible seriously is no simple matter. As a practical step in our quest to discover what the Bible teaches us concerning the starry heavens, however, I would like to suggest that taking the Bible seriously involves the following four actions: (1) affirming its true status, (2) respecting its multifaceted character, (3) promoting its proper function, and (4) engaging in a disciplined study of what it has to say.

THE STATUS OF THE BIBLE

What is the *status* of the Bible? Where does it stand relative to other books? What is its place relative to the entire body of human literature? It is crucial that we resolve the question of status, because it affects in a significant way nearly every other question we might ask concerning the nature of the Bible.

Stated in rather simple form, the Bible has the status of "holy Scripture"—the "Word of God" expressed in the form of human language and literature. To see more clearly how these serve as statements about the Bible's status, note how their traditional understanding leads to a determination of the Bible's place, or standing, relative to other literature.[1] As the Word of

1. When I say "traditional understanding," I should indicate the particular line in which I stand. Though I feel no obligation to defend any given individual's formulation of a statement concerning the status of the Bible as the Word of God, I suspect that my own statements will reveal that I was trained in a Calvinist tradition, largely rooted in Augustinian teaching, and that my particular branch

God, the Bible stands in a unique, elevated position relative to all of human literature, being neither totally within the category of human literature, nor entirely outside of it.

The Bible is more than a collection of ancient expressions of human religious experience; its roots reach far beneath the surface of such experience into the depths of divine revelation. But that is not to say that it stands entirely outside the category of human literature. Though it has the *status* of the "Word of God," the Bible comes to us in the *form* of thoroughly human language and literature.

The Bible is the "Word" of God, not the "words" of God. The Bible did not drop from the sky by an act of divine magic. God did not circumvent human means of writing, editing, and assembling the body of legal, historical, and literary documents that constitute the Bible. Yet, while the words of the Bible were produced by human writers, the Bible as an organic whole functions as God's Word, holy Scripture.

We should, therefore, be alert to two ways in which the status of the Bible is often incorrectly identified. There is on the one hand the error of placing the Bible entirely within the category of human literature and on an equal standing with it—an error common among persons whose worldviews draw heavily on philosophical materialism, naturalism, or humanism. There is on the other hand the error of placing the Bible entirely outside the category of human literature as if it were divinely dictated to mechanical printing machines. Such an approach, and others closely related to it, lead to the all too common phenomenon of breaking the Bible into many separate pieces, which, when isolated from one another, or isolated from their cultural, historical, literary, and canonical contexts, can be forced to support all manner of bizarre speculations (as we will see later).

The true status of the Bible, then, is properly identified by the phrase "Word of God." This clearly indicates that it occupies an elevated position relative to other human literature. And if we understand that the term *Word* is being used in a metaphorical sense to acknowledge divine revelation, rather than in the restricted literal sense to indicate mere words, then we can also avoid the error of denying the form in which God has chosen to reveal himself to us.

---

of the Reformed tradition is represented by such men as Abraham Kuyper, Herman Bavinck, and G. C. Berkouwer in the Netherlands, and Louis Berkhof in America.

We can shed additional light on the status of the Bible by saying that it has the status of "covenantal canon," an aspect that involves its functional authority in our lives. In this context, the word *canon* designates a body of documents with regulative and binding authority, and the word *covenant* designates a formal agreement between two parties not necessarily of equal status.

The unifying theme of biblical history, from Sinai to the ascension, is God's promise (his covenant) to be our faithful Lord and our responsibility to be his faithful servants. That covenantal character of history has, in turn, led to the covenantal character of biblical canon. As Meredith G. Kline has pointed out,

> The origin of the Old Testament canon coincided with the founding of the kingdom of Israel by covenant at Sinai. The very treaty that formally established the Israelite theocracy was itself the beginning and the nucleus of the total covenantal cluster of writings which constitutes the Old Testament canon. . . .
>
> In the case of the New Testament as in that of the Old Testament, acceptance of its own claims as to its primary divine authorship leads to recognition of its pervasively covenantal nature and purpose. For the New Testament so received will be understood as the word of the ascended Lord of the new covenant, by which he structures the community of the new covenant and orders the faith and life of his servant people in their consecrated relationship to him.[2]

Kline also reminds us that the words *covenant* and *testament* are nearly equivalent terms and that the traditional titles of "Old Testament" and "New Testament" are appropriate designations not only of the character of the Bible's contents but also of their status as legal documents: "The documents which combine to form the Bible are in their very nature—a legal sort of nature, it turns out—covenantal. In short, the Bible *is* the old and new covenants. . . . All Scripture is covenantal, and the canonicity of all Scripture is covenantal. Biblical canon is covenantal canon."[3]

Kline's contention that biblical canonicity is covenantal in nature is not only important to us in our efforts to affirm the true status of the Bible but also helpful to us in our efforts to understand the *character* of Scripture and to appreciate its multifaceted nature.

---

2. Kline, *The Structure of Biblical Authority*, 2d ed. (Grand Rapids: Eerdmans, 1975), pp. 43, 71.

3. Kline, *The Structure of Biblical Authority*, p. 75.

THE MULTIFACETED CHARACTER OF SCRIPTURE

Though it is unified as the covenantal, authoritative Word of God, the Bible is also rich in the diversity of its contents, sources, and forms. I will make no attempt to provide an exhaustive account of this variety, but we should note a few relevant points as a basis for later reference.

The Diversity of Scripture's Content

The Bible is filled with direct declarations that there is only one God who is the sovereign Creator of all things and persons. Inseparably linked with these declarations are numerous promises that our Creator-God is also our Redeemer-God. In the fullness of the revelation of the New Testament (new covenant), we are shown not only that creation and redemption are inseparably linked but also that they are rooted in and culminate in the work of Jesus Christ.

Scripture also provides us with ample testimony to God's faithfulness as historically experienced in the lives of his people. A multitude of witnesses to God's mighty acts attest to his majesty, power, righteousness, and goodness. The pages of Scripture are filled with their praise of the Creator-Redeemer.

But the biblical covenant also establishes that the servants of the Creator-Redeemer have certain obligations. Scripture provides us with instruction concerning our faith (what we are to believe concerning God and his promises) and our life (how we are to live in gratitude for his redemptive acts).

And finally it must be noted that much of what we find in many parts of the Bible is merely incidental information of little importance or relevance to its gospel message. I open my Bible randomly and happen on 1 Chronicles 20:6, for instance, in which I am informed that there was a man from Gath who had six fingers on each hand and six toes on each foot, twenty-four digits in all. Now that's interesting, but not particularly important or relevant to my redemption.

The Multiplicity of Scripture's Sources

In discussing the status of Scripture, we noted that the Bible is neither exclusively human nor exclusively divine. One way to understand this unique status is to investigate the sources the writers of Scripture drew on as they formulated and assembled the documents we now call the Bible. I am not thinking narrowly of just the literary or traditional sources upon which the present text is based but rather of two broader categories of source: divine revelation and human experience. It has long been one of the basic tenets of the Christian faith that God is the ultimate author of Scripture, that Scripture is "God-breathed." In a real sense, what Scripture tells us about God is not merely a compilation of mankind's ideas about divinity but is rather the result of divinely initiated revelation. It is God who took the initiative, God who guided the development of perceptions concerning his being and our relationship to him, God who provided individuals and communities with the knowledge and skills needed to commit these perceptions to writing, God who directed their writing so that the redemptive message carried by Scripture is the message of God without failure.

Yet, while the content of the Bible's kerygmatic message, its proclamation of the gospel, is thoroughly divine because of its origins in divine revelation, the form of that message and the historical-cultural context in which it was revealed is thoroughly human. Though the taproot of Scripture reaches into the depths of divine revelation, the lateral roots extend into the soil of human experience.

Both sources, human and divine, are necessary for Scripture to be what it is. Divine revelation is necessary in order that Scripture have authority. The declarations and promises of God found in the Bible have meaning only if their source is divine revelation: only God can speak for himself. Similarly, the Bible's requirements for our faith and life have meaning only if they are rooted in divine revelation; only God has the authority to make such demands on us. At the same time, the reality of the human experience recounted in Scripture is necessary for its authenticity. The human experiences with the material world recounted in the Bible are authentic experiences. The human history recorded in Scripture is genuine human history. The incidents of human interaction with God attested to in the Bible are instances of genuine human-divine encounters. The ring of authentic human experience is heard throughout the entirety of Scripture.

But how, one might ask, is it possible for the Bible to be both thoroughly divine and thoroughly human? For centuries theologians have struggled to understand the mechanism of divine inspiration. Such terms as "verbal inspiration" and "organic inspiration" and "plenary inspiration" will be found in the abundant literature on this subject. I have no desire to attempt a review of the technicalities of such distinctions, nor am I convinced that the heart of the matter lies in these technical differentiations. Such a procedure can easily degenerate into a coldly mechanical approach to a phenomenon the essence of which lies beyond the limits of mere mechanism. It is like trying to describe the performance of a ballet dancer in terms of force vectors, accelerations, and displacements—the language of Newtonian physics; accurate, perhaps, but missing the essential qualities of the performance.

A far better approach, it seems to me, is to draw the traditional analogy between the divine Word expressed in human words, the Scriptures, and the divine Word manifested in human flesh, the Christ.[4] The prologue to the Gospel of John expresses it so magnificently:

> The Word was made flesh,
> he lived among us,
> and we saw his glory,
> the glory that is his as the only Son of the Father,
> full of grace and truth. (John 1:14)

Just as we have difficulty in fully comprehending the incarnation of God's Son in human flesh, so too in the inscripturation of God's Word in human language we encounter a mystery beyond human understanding. Yet, while the mechanism of such an integration of human words and the divine Word lies outside of our comprehension, the reality is there to be seen and appreciated for what it is.

The Variety of Scripture's Forms

Those in the mainstream of the Judeo-Christian tradition have never considered the Scriptures to be merely ancient religious literature. To say that Scripture is nothing more than religious literature is to fail to affirm its true status as the Word of God. It *is* literature, however, and it must be respectfully treated as such. As Leland Ryken says, "Christianity is the most literary religion in the world and the one in which the word has a special sanctity. The clearest evidence of this literary emphasis is the Bible, which is not only the repository of Hebraic-Christian belief but

4. See G. C. Berkouwer, *Holy Scripture,* trans. and ed. Jack B. Rogers, Studies in Dogmatics (Grand Rapids: Eerdmans, 1975), pp. 195-212.

is also a book on which literary form is of overriding importance." Moreover, "The artistic beauty of the Bible exists for the reader's enjoyment and artistic enrichment. To ignore this aspect of biblical literature is to distort the Bible as a written document."[5]

Why are the Scriptures' literary qualities noteworthy? First, it is essential to serious biblical study to become aware of the overall structure of the Bible. The Bible is not merely a collection of divinely inspired documents randomly assembled; it has been assembled according to a definite pattern. Biblical scholars have presented strong evidence that the overarching structure of the Bible is the structure of the suzerainty covenant—a legal and literary form of treaty (covenant) administration that was well established in the ancient Near East at the time the Old Testament was written. Thus it appears that Scripture has not only the *status* of covenant, but also the *form* of covenant as well. As Kline puts it, "whatever the individual names of the several major literary genres of the Old Testament, as adopted in the Old Testament their common surname is Covenant."[6] Therefore when we seek to understand any particular part of Scripture, it is essential that we consider the role or function of that part in the context of the whole covenantal structure of the Bible. Such consideration will be particularly helpful in our later discussion of Genesis 1.

Literary considerations are also important when we seek to identify the literary form of individual segments of Scripture. The Bible is overwhelmingly rich in the variety of genres it contains. Leland Ryken cites the following literary forms: "the story of origins, heroic narrative, epic, parody, tragedy, lyric, epithalamion, encomium, wisdom literature, proverb, parable, pastoral, satire, prophecy, gospel, epistle, oratory, and apocalypse."[7] Gerhard Lohfink cites a different set of genres: "historical account, saga, myth, fairy tale, fable, paradigm, homily, admonition, confession, instructive narrative, similitude, parable, illustrative saying, prophetic utterance, juridical saying, wise saying, proverb, riddle, speech, contract, list, prayer, song."[8]

---

5. Ryken, *The Literature of the Bible* (Grand Rapids: Zondervan, 1974), pp. 9, 14.

6. Kline, *The Structure of Biblical Authority*, p. 47.

7. Ryken, *The Literature of the Bible*, pp. 14-15.

8. Lohfink, *The Bible: Now I Get It!* (Garden City, N.Y.: Doubleday, 1979), p. 64.

Is it necessary or even important for the Bible reader to be aware of this variety of genres? Clearly it is, because each form conveys its message in a unique manner; the correct meaning of any particular passage can be accurately determined only by the use of an appropriate program of interpretation. Different literary genres call for different methods of interpretation. Failing to recognize a given literary form correctly will most likely make it impossible to determine the intended meaning of any kind of literature, biblical literature included. If we try to read lyric poetry as if it were descriptive prose, we might find preposterous absurdity rather than a beautiful and imaginative representation of a profound truth. If we try to read heroic narrative as if it were historical chronicle, we might get an unrealistic or distorted picture of the nature of humankind and our interaction with God. If we try to read a poetic or liturgical story of origins as if it were a primitive scientific report, we might see a chronicle of divine magic rather than an artistic portrait of the Creator-Creation relationship.

In all cases, the failure to identify the genre of a piece of literature correctly not only prevents us from getting its real meaning but also may lead us into extracting a distorted, perhaps even grotesque, misinterpretation in its place. The loss is thereby doubled!

Twentieth-century Western culture seems to me particularly inept at understanding and using figurative or symbolic literature. We are so accustomed to straightforward, matter-of-fact descriptive prose that we expect nearly all writing to be of that form. While I should be one of the last persons to underrate the positive contributions of modern science and technology to Western culture, yet I would be one of the first to point out some of its negative contributions as well—one of which is an apparent dulling of our awareness of both the need and the power of symbolism, imagery, and all of the other artistic elements of figurative literature. Expository scientific writing has made an illegitimate claim of supremacy over artistic literature. The technical journal article, though its author may vehemently protest, has no more claim to truth than does the poem. Perhaps less.

Ignorance and ineptness are often accompanied by fear. Fear of the unfamiliar may be the most common kind of fear. Many sincere Christians who wish to take the Scriptures seriously take comfortable refuge in a literalistic reading of nearly all of Scripture because of their fear of venturing along the less familiar path of a literary reading, particularly if it leads through the regions of symbolic or figurative interpretation. There is a genuine fear that in the course of dealing with the Bible literarily we may lose hold of its surety, its fundamental doctrines, its changeless truths. If you treat this part of Scripture figuratively, they say, then soon everything will be no more than symbol and image, leaving nothing of substance or value. The doctors of poetry will surgically remove the heart of the Bible and replace it with a flower. It may look nice, but the patient will be dead.

I must insist that such a fear is not only unfounded but also untrusting; it displays a lack of trust in Scripture's author. If we really have faith that God is the ultimate author of Scripture, then we will be confident that he has guided his agent-writers in such a way that the product, his Word in human language and literature, will stand up to any legitimate test. Since God's Word is carried by the vehicle of human literature, we can be sure that the application of the best principles of literary analysis will never destroy or damage his message. On the contrary, we may confidently expect that we will see his message all the more clearly when we make proper use of critical tools.

THE FUNCTIONS OF SCRIPTURE

As I suggested, taking the Bible seriously entails promoting its proper function. In a general way, we can say that the Bible will function properly only when it is being used according to its intended purposes. In considering the status and character of Scripture, we necessarily came across some references to its functions or purposes, but let us now assemble these and other related matters into a brief review of how Scripture must function in the believer's life of faith and action.

The primary purpose of Scripture is, in my judgment, to bring the reader into a right relationship with God, with fellow human beings, and with the rest of the Creation. This is the kerygmatic heart of Scripture's proper function—the bringing of the gospel message. Functioning in this way, the Bible provides for believers both the elements of Christian faith and the principles for Christian life. Most importantly, it is by this means that a person meets God personally. Through the testimony of the Holy Spirit, the words of the Bible come alive as the Word of God, enabling the believer to hear the covenant promises from God himself, to hear the call to belief, obedience, and love from the Master himself.

Having personally met the Creator-Redeemer, the believer

is in a position to hear the testimony and accept the witness of many other believers who have experienced God's mighty acts in history and in their own lives. It is a secondary purpose of the Bible to provide the believer with such testimonies and eyewitness accounts of God at work in human history and in his Creation. These are not offered as "proofs" of the veracity of God's claims; God's Word to mankind needs no proof. Rather, their incorporation into Scripture is evidence of the authenticity of the human experience with the dynamic divine presence in the world and in its history.

Yet another function of Scripture, of considerably less importance than the primary and secondary functions, is to provide the reader with information concerning things in themselves. I say "in themselves" as opposed to "in relation to God" or "in their function as vehicles of divine revelation." Such information is incidental to the principal themes of Scripture and is drawn solely from human experience. This includes information about persons (such as the man from Gath who had twenty-four digits), information about events, or information about the material world expressed in the pre-scientific language of the day. Such information may be relevant for academic study of one sort or another but makes little or no impact on our experience of redemption. Taking the Bible seriously does not require us to treat such matters as the product of divine revelation; on the contrary, taking the Bible seriously requires, I believe, giving such incidental information the lesser status that I have suggested.

THE DISCIPLINED STUDY OF SCRIPTURE

In the context of what we have noted about Scripture, it should be quite obvious that the Bible deserves careful and disciplined study. In that study we must establish certain priorities. Our first priority should be to study the Bible as Scripture, as God's covenantal Word. By this means we come to hear God's declarations and promises and to learn of our responsibilities for living in thankful obedience. A second and closely related priority is to read the Bible as a witness and testimony to God's acts in the lives of communities and individuals in the past. And, finally, we may also study the Bible as a source of information concerning historical events and persons, ancient Near Eastern cultural phenomena, and the like.

While the Bible's central message of redemption is sufficiently clear for any person to understand, we must nevertheless recognize that all academic disciplines may be profitably employed to gain a deeper and more detailed knowledge of God, his people, and his world. For this purpose we read and study the Bible in a disciplined, analytical, or "critical" fashion in order better to know such things as the text of the Bible, the historical and cultural contexts in which it was written, the literary forms of Scripture, and the history of the formation of the biblical canon.[9] It must be kept in mind, however, that all of these analytical studies must be employed in service to the primary covenantal function of Scripture; outside of this service they are of quite limited value.

The Vehicle Model of the Bible

Scripture is often referred to as a vehicle of God's communication to humanity. This is an apt metaphor, pointing to the fact that indeed if there is to be a message carried or conveyed from God to mankind, then there must be some concrete, or at least identifiable, means of conveyance: there must be a vehicle that transports the message from sender to recipient.

But let us press the vehicle metaphor a bit further. Just as in the world of commerce many types of vehicles are used to convey goods from producer to consumer, so too there are many types of literature, many literary genres, that God uses in the Scriptures to convey his message to his audience, his people. As the type of vehicle is chosen according to the nature of the goods to be carried, so in the Bible the literary genre is suited to the message it must convey. Simple matters of historical record may be conveyed by a matter-of-fact chronicle of events. Profound truths of immense magnitude, however, cannot always be adequately expressed in the genre of straightforward expository discourse; they are often expressed better in a more symbolic or poetic form. How often we say, "Words just can't describe what I want to express." Our best alternative, then, is to shift from expository discourse, which does constitute an attempt to contain something in words fully, to poetry or some other highly symbolic form that makes no pretense of exhaustively describing some great thought or event or emotion, but instead freely admits that it is simply pointing in a certain direction that readers

9. For an introduction to the positive aspects of the term *criticism*, see Harry R. Boer's *The Bible and Higher Criticism* (Grand Rapids: Eerdmans, 1981).

must creatively and imaginatively follow if they are to get even the beginning of an understanding of that profound idea.

Because we find many forms of literature in the Scriptures, we would do better to think of them not as a single vehicle but rather as a caravan of various vehicles, each suited to the task of conveying a particular message from God to man.

To press the model still further, we might note that when goods are carried by a vehicle, it is wise to package the goods appropriately in order to protect the contents from damage and to provide convenient units for handling and delivery. Similarly in the Scriptures, each vehicle is loaded with its content (God's message) contained in appropriate packaging—the specific story or account of an event; the particular symbolism used in a poem; the specific cultural patterns that form the context of commentary or instruction or description.

Consider Psalm 23, for example. The vehicle employed here is poetry—poetry serving as a vehicle to convey to us the message that God loves us, cares for us, and provides abundantly for all of our needs. This message of love and care, conveyed by the vehicle of lyric poetry, is packaged in pastoral language—the metaphor of a benevolent shepherd caring for his sheep. In such an appropriate package the message can be effectively conveyed to readers of all ages and all times. The magnitude of that message is so great that only the vehicle of lyric poetry could bear the heavy load. Powerful vehicle, appropriate packaging, magnificent message!

To complete the introduction to this vehicle model of Scripture, let me note, finally, that the Bible can be viewed as a complete unit, including the vehicle (literary genre), packaging (specific story, symbols, etc.), and contents (God's message to us). The message or *teachings* of the Bible come from only one source—God. Scripture ought never to be viewed as a mixture of God's teachings and man's teachings. Since all of the teachings of Scripture come from God they are trustworthy and authoritive. We can be confident that all of the content of God's message in the Bible is delivered to us undamaged and unspoiled. It ought never to be viewed as a mixture of teachings, some true and others false. However, as our model suggests, just as a consumer must first unload the packaged goods from a delivery vehicle and then carefully unpackage those goods for use or consumption, so we as readers of Scripture must be studiously and prayerfully wise in separating the contents (the trustworthy teachings of God) from the vehicle and the packaging. Neglecting that separation would be as foolish as attempting to eat a granola bar without first removing it from its wrapper or, more absurd yet, without distinguishing it from the truck that delivered it to the store.

Before we leave this discussion on the structure and function of Scripture, I would like to look at this topic through the eyes of one with far greater vision than I would claim for myself. Without doubt, C. S. Lewis is one of the most widely respected writers of Christian literature in the twentieth century. He possessed a rare and priceless combination of literary creativity, human perceptiveness, and Christian commitment that could not fail to touch a reader. My father, having emigrated from the Netherlands to America when he was twelve years old, had no opportunity for formal education beyond the sixth grade. He did, however, have a lifelong thirst for learning. He learned much by regular reading of good literature, and C. S. Lewis was clearly his favorite writer. I now have on my bookshelf many of Lewis's works from my father's collection. They are treasures, both for the memory of my father and for the rich Christian insights so creatively expressed in them.

In *Reflections on the Psalms,* Lewis discusses his views on the nature and function of Scripture, making frequent use of the vehicle metaphor. His development of that metaphor, however, is different from mine, placing more emphasis on divine use than on divine origin or content. I would like to summarize his approach here. Some readers may prefer it to mine.

Lewis's picture of Scripture is not one of God's Word "sent down" from heaven in the guise of human literature but rather one of human literature being "taken up" into God's higher service. The Scriptures are "holy" not because of extraterrestrial origins, he suggests, but for their heavenly function. Speaking of the poetic story of origins found in Genesis (of which I will say more in subsequent chapters), Lewis comments,

> something originally merely natural—the kind of myth that is found among most nations—will have been raised by God above itself, qualified by Him and compelled by Him to serve purposes which of itself it would not have served. Generalizing this, I take it that the whole Old Testament consists of the same sort of material as any other literature—chronicle (some of it obviously pretty accurate), poems, moral and political diatribes, romances, and what not; but all taken into the service of God's word.[10]

10. Lewis, *Reflections on the Psalms* (New York: Harcourt Brace Jovanovich, 1958), p. 111.

And what are the consequences of this "taking up" process? What do we learn from Lewis concerning the nature of the product—Scripture?

> The total result is not "the Word of God" in the sense that every passage, in itself, gives impeccable science or history. It carries the Word of God; and we (under grace with attention to tradition and to interpreters wiser than ourselves, and with the use of such intelligence and learning as we may have) receive that word from it not by using it as an encyclopedia or an encyclical but by steeping ourselves in its tone or temper and so learning its overall message.[11]

Lewis goes on to comment that, had we been consulted, we might have requested that God provide us with quite a different sort of document, "giving us ultimate truth in systematic form—something we could have tabulated and memorized and relied on like the multiplication table." But even the recorded teachings of Jesus cannot be walled in by the unyielding bricks of philosophical proposition and logical analysis. Why were we given something radically different? Because God knows our needs better than we.

> It may be indispensable that our Lord's teaching, by that elusiveness (to our systematizing intellect), should demand a response from the whole man, should make it so clear that there is no question of learning a subject but steeping ourselves in a Personality, acquiring a new outlook and temper, breathing a new atmosphere, suffering Him, in His own way, to rebuild in us the defaced image of Himself.[12]

In other words, we have, by God's wisdom, been given something that demands our total response, not just our intellectual response. Or, to express that in the terminology of my Calvinist heritage, we must live our whole lives with our whole being as in the presence of the living sovereign God. We belong not to the Idea but to the Person.

The Task of Interpreting Scripture

Our disciplined study of the Bible must be guided by sound principles of interpretation. The study of such principles themselves is called *hermeneutics,* and the application of these princi-

---

11. Lewis, *Reflections on the Psalms,* p. 112.
12. Lewis, *Reflections on the Psalms,* pp. 113-14.

ples to determine the meaning of a particular message is called *exegesis*. Drawing from our vehicular model of the Bible, I would like here to suggest a "vehicle-packaging-content" approach to biblical interpretation. Clearly this will not be a highly structured and detailed development such as a professional theologian would offer; rather, my intent is to offer a picture of how these more sophisticated techniques function in practice—an approach more easily visualized and used by amateurs.

If we are correct in supposing that the Scriptures represent a unified combination of vehicle, packaging, and contents, then the first operation of biblical interpretation must be to distinguish among the three categories. The next will be to extract the contents from both the vehicle and the packaging. Or, to put it in "real" terms instead of "model" terms, the task of biblical intepretation is to extract the original meaning, God's message or teaching, from the specific event, account, or story as it has been conveyed to us by a particular literary genre, such as chronicle, epic, or parable.

All of the traditional principles of biblical hermeneutics must be employed in these operations. We must, for instance, employ the principles of linguistic analysis and literary criticism to identify correctly the literary genre, the grammatical constructions, and the specific meaning of each word in order to perceive the theme or situation in which and for which a passage was written. We must be thoroughly familiar with the cultural settings of a passage to ascertain what its original readers would take it to mean. As Berkhof suggests, the exegete "must place himself on the standpoint of the author, and seek to enter into his very soul, until he, as it were, lives his life and thinks his thoughts. This means that he will have to guard carefully against the rather common mistake of transferring the author to the present day and making him speak the language of the twentieth century."[13] Finally, we must interpret any Scripture passage in the context of the primary function of Scripture—to reveal God's marvelous work of redemption. When we read Scripture we have to realize that it is not primarily a treatise on philosophy, though philosophical principles may be drawn from it; that it is not primarily a chronicle of history, though its message is rooted in historical events; that it is not primarily a textbook of natural science, though it contains important teachings about the

---

13. Berkhof, *Principles of Biblical Interpretation* (Grand Rapids: Baker Book, 1950), p. 115.

material world. We will do well to keep all of these matters in mind as we proceed to search the Scriptures for the particular purpose of learning what they teach about the world of sun, stars, and galaxies and compare that with what we learn about those same celestial bodies through astronomical observation.

CHAPTER TWO

The View from Palestine

Experience teaches us that the appearance of the starry heavens depends very directly on our vantage point: what we see on the celestial sphere is determined by the location from which we view it and by the time at which we do our viewing. The sky as viewed from London appears quite different from that observed from Lima. The constellations seen in December are not the same ones that we see in July. These differences arise entirely as a result of geometrical relationships and are the topic of thorough study in positional astronomy.

But there is another very important way in which vantage point controls what we perceive in an upward gaze: our cultural heritage and historical context influence how we mentally interpret or understand what we visually discern. Because of our particular training, we twentieth-century Westerners assign a meaning to astral objects and celestial phenomena quite different from the meaning given to those same objects and phenomena by people who lived in Palestine three thousand years ago.

In light of this, if we truly wish to discover what we might (or might not) learn about celestial luminaries from the Bible, we must spend some time in preparation. Just as vigorous athletic activity should be preceded by stretching and warm-up exercises, so also serious Bible study ought to be preceded by mental exercises that stretch our awareness and warm up our intellectual curiosity. In preparation for viewing the celestial sphere through the lens of Scripture, we should take a few moments to familiarize ourselves with the historical and cultural context in which the Bible was written, particularly the context in which the Old Testament teaching concerning the stars as God's Creation was formulated.

THE BIRTH AND DEVELOPMENT OF THE NATION ISRAEL

Palestine has been the site of human civilization for several millennia. Archeological evidence indicates, for example, that the region near the city of Jericho was occupied as long ago as 8000 B.C. But the region that appears to have given birth to the first highly developed human culture is known as Mesopotamia—so named because it lies primarily between the rivers Tigris and Euphrates. This is the area now occupied by the country of Iraq. Already in the third and fourth millennia B.C., numerous societies flourished in this territory, populated by such groups as the Sumerians, Akkadians, Elamites, Kassites, and Hittites. Cities of importance in this era (some of which were city-states, others the chief city of a dynasty) include Uruk, Lagesh, Erech, Susa, and Ur. Culture and technology experienced very significant development during this period. The Sumerians, for example, are credited with the development of writing, even before 3000 B.C. The ancient Mesopotamians made tools and decorative objects of bronze and copper with great skill.

Though Israel did not attain the status of a nation until after the Exodus, its origin is commonly assigned to the much earlier period of the patriarchs. Sometime early in the second millennium B.C., Abraham and his clan migrated from northern Mesopotamia (probably the area near the ancient city of Haran) to the region of Palestine. The Bible, drawing from the memory of the Hebrews as it was preserved primarily in oral tradition, provides us with sketches of some events and individuals of the patriarchal period but is far more concerned with providing religious history and with presenting a testimony to the reality of divine action and leading. Specific dates are not available for either the life of Abraham or his entry into Egypt, although this event is generally agreed to have occurred before the middle of the second millenium, perhaps around 1700 B.C.

If the conception of Israel is identified with the migration of the patriarchs from Mesopotamia to Palestine, the birth of Israel would surely be identified with Moses and the Exodus—the divinely ordained delivery of the oppressed Israelites from

Egypt into the promised land of Canaan. Another event as significant as the Exodus took place as Moses and the Israelites traveled from Egypt to Canaan: the covenant was established at Sinai. Yahweh promised that he would be Israel's faithful God, and Israel promised to serve Yahweh obediently in gratitude for his goodness to them, as evidenced by their delivery from bondage in Egypt.

The Exodus is commonly dated around 1280 B.C. and the Israelite presence in Canaan was firmly established by around 1200 B.C. This marked the beginning of the period of the "judges," which spanned the twelfth and eleventh centuries B.C. and was a period of considerable turmoil and conflict. During this period Israel had no centralized government; temporary leadership arose in response to specific needs or threats, such as the continual challenge presented by the Philistines (from whom the name "Palestine" is derived).

The loose confederacy was soon replaced by a strong monarchy. Israel became a united people—a nation with a king. The "golden age" of Israel occurred under the kingship of Saul (ca. 1020-1000 B.C.), David (ca. 1000-961 B.C.), and Solomon (ca. 961-922 B.C.). But following that period of unity and prosperity the kingdom was divided in 922 B.C. into a northern faction (Israel) and a southern faction (Judah).

The Northern Kingdom of Israel, the capital of which was first located at Shechem and later at Samaria, lasted just two hundred years. During that time it experienced periods of civil strife, prosperity, and decline. It vacillated between friendship and enmity with Judah. It contended with numerous external powers, such as Syria, Assyria, and Egypt. During that period the people of Israel repeatedly yielded to the temptation to abandon their covenantal promise to obey Yahweh and chose instead to serve Baal and the other gods of Canaanite paganism. Baalism was strongly promoted by Jezebel, wife of the unscrupulous King Ahab. In the end it was Assyria who shattered Israel, which was forced to surrender its capital, Samaria, to Sargon II in 721 B.C.

Judah, to the south, survived for an additional century and a third. Like the Northern Kingdom of Israel, Judah also went alternately through periods of turmoil and prosperity. Battles with Israel, Assyria, Egypt, and Babylon sapped its strength and resources. Athaliah, daughter of Jezebel and Ahab from the north, became the wife of Judah's King Jehoram, and with her came the strong influence of pagan Baalism. Some time later,

King Manasseh encouraged the worship of celestial deities, much like his neighbors, the Assyrians. Josiah attempted to turn Judah back onto the track of covenant faithfulness to Yahweh, but his efforts were only partially successful. The rising power of Babylon was ultimately overwhelming, and in 587 B.C. Judah's capital city of Jerusalem fell to Nebuchadnezzar. Jerusalem and its temple were virtually destroyed; the people were widely scattered, with a large contingent of its leading citizens carried into exile in the city of Babylon.

While exiled in Babylon, the Jews (the people of Judah) became fully acquainted with the prosperous and highly developed culture of the Babylonians. Yet, in the midst of this thoroughly pagan polytheistic society, the Jews longed to see the restoration of the nation of Israel—the people of Yahweh. That dream was soon to come true. In 539 B.C., Cyrus the Persian captured Babylon, and a year later gave the Jews permission to return to their own territory.

Although the temple was rebuilt by 515 B.C., it was nearly another century before Jerusalem's walls of fortification were rebuilt under Nehemiah's leadership. Many Jews were still scattered in such distant areas as Egypt and Babylonia. But a restored community of Israel was developing near Jerusalem. Ezra led that community in a magnificent ceremony signifying the renewal of its covenant with Yahweh, reminiscent of a similar covenant renewal under Josiah some two centuries earlier.

Thus Israel was reborn. As it had been delivered from slavery in Egypt, it was now delivered from captivity in Babylon. As it had once entered into a sacred covenant with Yahweh at Sinai, so now that covenant relationship with Yahweh, its Creator and Redeemer, was restored.

Political oppression, however, did not cease. Palestine, along with its larger and more powerful neighbors Persia and Egypt, fell to Alexander the Great of Greece around 332 B.C. When Alexander died, his kingdom was divided, leaving Palestine first in the hands of the Ptolemies of Egypt until 198 B.C. and then under the power of the Selucid dynasty, which also ruled Babylon.

While the Ptolemaic rule was relatively tolerant of the Jewish religion, the Selucid rulers were determined to spread the Greek culture they now represented. Antiochus Epiphanes tried his best to stamp out Judaism and to replace it with the worship of the Greek god Zeus. Many Jews rejected this attempted Hellenization and organized a strong resistance under the leadership of Judas Maccabeus. The Maccabean wars provided the Jews with nearly a century of relative independence until Palestine came under the rule of the Roman Empire in 63 B.C.

This is of course only a thumbnail historical sketch of the period,[1] but it does point to the fact that the people of Israel were, during the entire Old Testament period, in times of both peace and war, in very close contact with their neighboring cultures. This would suggest that in attempting to understand how the Jews perceived the celestial luminaries (and the material world in general), we will do well to compare and contrast Old Testament references to the celestial realm with the concepts held by Israel's neighbors in Mesopotamia, Egypt, and Canaan (Palestine). Before we look at the Hebrew concept of the heavens, then, let's review some of the developments in early technical astronomy and some aspects of ancient Near Eastern astral religious belief.

ANCIENT PRACTICAL ASTRONOMY: CELESTIAL TIMEKEEPERS

The vast majority of commonly used time intervals are astronomical in origin. The day, to pick the most obvious, is established by the cyclic appearance of the sun above the horizon, providing alternate periods of light and darkness. The month, as the name itself implies, is based on the periodic phases of the moon. The visible portion of its illuminated surface waxes (grows) from the barely visible crescent—which the ancients called the "new moon"—to the fully illuminated disk of the full moon, and then wanes to invisibility, soon to reappear as a thin crescent. The average time interval between successive new moons is about 29.5 days. Therefore, ancient calendars commonly used months of either 29 or 30 days. The year, the time for one complete cycle of seasons, is established by the apparent motion of the sun, though in a somewhat more complex fashion, as we shall soon see.

Smaller time intervals are simply fractions of a day—not fractions chosen for convenience according to modern computational procedures but rather fractions that represent vestiges of ancient numerical conventions. Perhaps as early as 2000 B.C. the

1. For a thorough treatment of this topic, see John Bright, *A History of Israel,* 3d ed. (Philadelphia: Westminster Press, 1981).

Egyptians chose to divide both day and night into twelve units. Since the length of daylight and darkness vary in the course of the year, these "seasonal hours" also varied in length, which made the design and construction of mechanical clocks quite difficult. Later, in the Hellenistic period, these seasonally varying hours were replaced by "equinoctial hours," hours of constant length, twenty-four per solar day.

The division of hours into smaller units follows the Babylonian number system, which is based on the number sixty. Hours are divided into sixty minutes, minutes are divided into sixty seconds, seconds may be divided into sixty "thirds," and so on. We find a similar remnant of the Babylonian system in our convention of dividing a complete circle into 360 degrees of angle, with degrees divided into minutes and seconds of arc through successive divisions by sixty. All of these familiar units of time and angle that we use daily are remnants of numerical conventions that were established thousands of years ago in the ancient Near East.

The one time period in common use today that has a less obvious astronomical basis is the week of seven days. For the purposes of establishing and regulating cyclic periods of labor, commerce, or religious ritual, most cultures appear to have adopted a period of time between the day and the month. Beginning in the third millennium B.C., the Egyptian civil calendar was based on a year of 365 days—twelve months of thirty days each, plus five additional days. The month was further divided into three "decades" of ten days each. But evidence suggests that the Egyptians also divided the lunar month into periods of seven or eight days. The old Assyrian calendar may have divided the month into five periods of six days each. Various Mesopotamian calendars placed special emphasis on those days marking each quarter of a month. To this day we take note of the lunar cycle of phases at each quarter of the period: new moon, first quarter moon, full moon, and third quarter moon. To the nearest whole number, these phases are separated by seven days, suggesting that the convention of the seven-day week is tied to observation of the moon. Some scholars, however, hold that it is more likely that the choice of a seven-day calendric unit was based on a symbolic meaning assigned to the number seven than on the lunar cycle. In Israel, the liturgical calendar was firmly based on a strict seven-day cycle. The fact that such a cycle periodically gets out of synchronization with the lunar phases was considered a matter of little consequence. Perhaps Israel wished openly to defy association with the numerous pagan astral religions practiced by its neighbors, a matter we will touch on later in this chapter.

BABYLONIAN MATHEMATICAL ASTRONOMY

Quite appropriately, the Babylonians have been credited with the earliest development of a sophisticated computational astronomy. Old Babylonian texts indicate an interest in astronomical observation early in the second millennium B.C., concurrent with the patriarchal period in the history of Israel. These early texts, however, deal only with astrological omens and do not yet represent scientific astronomy. Texts dated around 700 B.C. show a far more detailed knowledge of astronomical phenomena but are still descriptive rather than computational in character. Mathematical astronomy appears to have begun in Babylon after that city was captured by the Persians in the middle of the first millennium. By two centuries later, when Alexander the Great swept through the area, it had been developed into a highly sophisticated system. While Israel was rebuilding its nation, its temple, and its capital city, the learned Babylonians, or Chaldeans as they are sometimes called, were becoming masters of astronomical computation.

Because astronomical observations in that era were unaided by precision measuring instruments (evidence suggests nothing but naked-eye observations), the reported positions of celestial objects are of very limited accuracy. The arithmetic computations of late Babylonian astronomy, however, demonstrate great cleverness and highly developed calculational programs. Before we look at a sample of Babylonian mathematical astronomy, though, we should first familiarize ourselves with certain celestial phenomena and a few technical terms.

If you or I were to look up into the sky on a clear evening, we would find it quite impossible to conclude much about how far away the many lights in the sky were located. The ordinary means by which we estimate distances to objects in our environment would fail to provide us with any values. We would be left, really, with little more to say than that everything must be very distant. It would be quite natural to suppose that all of the lights were equally distant from us, as if they were placed on the surface of an unimaginably large sphere surrounding us. And indeed, this is what the ancients believed. Though we have learned that this is not actually the case, the ancient concept of

Star trails formed by the Big Dipper (center) and the Little Dipper (top) around Polaris during a few minutes of rotation. Photos by the author.

the "celestial sphere" is still very helpful in describing the angular positions and angular motions of celestial bodies. If we are interested solely in directions and not in distances, the celestial sphere concept provides a very convenient aid in discussion.

Observing the celestial sphere for a time would soon reveal that it appears to move. Specifically, the celestial sphere appears to rotate once every 23 hours and 56 minutes around an axis that appears to pass very nearly through the star Polaris (the North Star). Just like the sun and moon, most stars appear to rise somewhere on the eastern horizon, trace parts of circular arcs westward across the sky, and set at corresponding points on the western horizon. Some stars near Polaris appear to trace out complete circular paths, leading us to refer to them as "circumpolar stars." All of the stars of the Big Dipper, for example, appear to be circumpolar stars when viewed from the northern U.S. and Canada.

Continued viewing of the celestial sphere over an extended time period would reveal even more. While the positions of the stars appear fixed on the celestial sphere, there are a few objects that appear to move relative to these fixed stars. The sun, the moon, and the planets all fall into this category. The word *planet,* in fact, is derived from the Greek word meaning "wanderer." Traditionally, planets are those celestial bodies that move, or wander, relative to the fixed stars on the celestial sphere. Modern usage of the term "planet" is restricted to those nine major objects—Mercury, Venus, the earth, Mars, Jupiter, Saturn, Uranus, Neptune, and Pluto—that orbit the sun.

The moon appears to move eastward relative to the celestial sphere at the rate of about thirteen degrees per day (though because the westward rotation of the celestial sphere is faster, the moon still appears to rise in the east and set in the west). Another way to describe the moon's apparent motion is to say that it appears to move eastward (relative to the celestial sphere, remember) through a distance equal to its own diameter in a time of one hour—something that we can easily observe on a clear evening.

If we could mark the sun's position relative to the celestial sphere, we would also observe its eastward motion, though it is considerably slower than the moon's. Each day, the sun's position relative to the celestial sphere (or, in modern terms, relative to the background of the much more distant stars) appears to move by one degree toward the east. As in the case of the moon, however, the more rapid westward motion of the celestial sphere makes the sun appear to rise from the eastern horizon, trace an arc westward across the sky, and set in the west. During the course of one year, the sun's apparent movement relative to the celestial sphere traces out a complete circular path that divides the celestial sphere into two equal parts. This "great circle" path is called the *ecliptic*.

Since the sixteenth century A.D. (the time of Copernicus), we have understood that the apparent motion of the sun along the ecliptic is the result of the earth's motion in its orbit around the sun. The Babylonians, however, did not ask concerning the causes of the celestial phenomena they observed. They carried out elaborate computational programs to describe certain aspects of planetary motions but developed no geometrical models or dynamical theories to explain why things appeared to move as they did.

The sun and moon appear to move on the celestial sphere along closely related circular paths at very nearly constant rates, but the movements of the planets on the celestial sphere are far more complex. The brightest of the planets, the ones visible to the ancients without the aid of telescopes, are Mercury, Venus, Mars, Jupiter, and Saturn. Though their respective motions differ in detail, they do share certain common features. Each planet appears to move eastward relative to the celestial sphere most of the time. We call this the planet's "direct" motion, and it takes place within a relatively narrow band, called the "zodiac," which is centered on the ecliptic. But this direct motion occurs at a varying rate, and at regular intervals (a different interval for each planet) a planet's direct motion comes to a halt. From that point, which the Babylonians called the "first stationary point," the planet proceeds to move westward relative to the celestial sphere for a time. This backward motion, or "retrograde" motion as it is traditionally called, first speeds up and then slows down again to a halt at the "second stationary point," from which it resumes its direct motion and repeats the whole cycle. The time for one complete cycle, called the planet's "synodic period," ranges from 116 days for Mercury to 780 days for Mars.

The path traced out by a planet on its apparent journey through the zodiac is certainly not a simple circle, as was the case for the sun and moon; rather, it contains loops or zig-zags formed by each retrograde excursion. A typical looping path is illustrated in figure 1.

Exercising great computational cleverness, the Babylonians prepared extensive tables of computed values for special

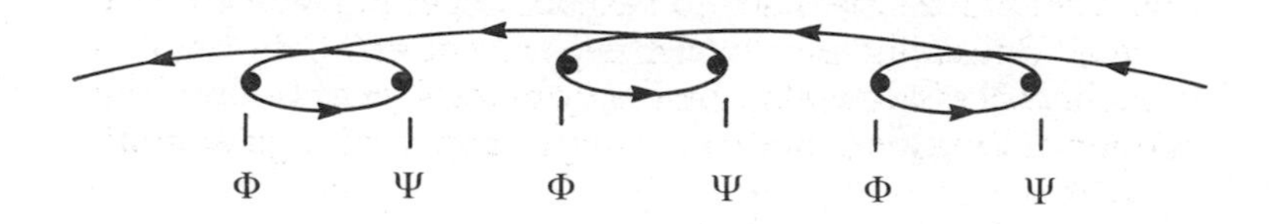

FIGURE 2-1. TYPICAL PATH OF A PLANET THROUGH THE ZODIAC. The points labeled Φ are first stationary points, while Ψ identifies the positions of second stationary points. Note how a planet's motion is predominantly direct (eastward), but that at regular intervals it moves in retrograde (westward) fashion between first and second stationary points.

positions and times related to planetary motion. Though the correspondence between actual and computed values is only approximate in any individual case, certain features of the Babylonian system display a remarkable precision. For example, according to the Babylonian tables (written in cuneiform on clay tablets), the average synodic period of Jupiter has a value of 398.89 days. Modern observation yields the value 398.88 days! We might be inclined to suspect that this is merely a coincidence, that the Babylonians made some lucky choices of parameters, but as it turns out, they managed to achieve the same precision in calculating the mean synodic period for all of the planets they observed. How did they obtain such remarkably good results without the benefit of precise measuring instruments? To this day no one knows.[2]

Though we may rightly stand in awe of the sophisticated computational astronomy of the Babylonians, we should also note its limitations. While it answered some questions with remarkable accuracy, it answered other questions not at all. More to the point, it failed even to ask several other questions that lie

2. For detailed discussion of the Babylonian computational programs, see O. Neugebauer, *The Exact Sciences in Antiquity*, 2d ed. (New York: Dover Publications, 1969). See also A. Aaboe, "Scientific Astronomy in Antiquity," *Philosophical Transactions of the Royal Society of London*, Pt. A, 276 (1974): 21-42.

at the heart of modern astronomy. Content with numerical programs to compute the times and locations of special points on a planet's path, Babylonian astronomy failed to develop a means of computing the planet's position at an arbitrary time and so could not predict the shape of the path traced out on the celestial sphere. Unlike its Greek counterpart, ancient Babylonian astronomy failed to develop geometric models for a planet's spatial motion as a means of understanding its apparent angular motion relative to the celestial sphere. Like all astronomy done before the work of Isaac Newton in the seventeenth century, it failed to deal with the whole question of the causes of planetary motion. This, I suggest, was not merely a matter of ignorance or of limited computational skills or even of crude observational methods, but rather a matter of cultural conditioning. Questions about material causes for the behavior of stars or planets were foreign to ancient thought. The ancients did not ask such questions because they would have made no sense within the world-picture or cosmology of their culture.

Pushing that consideration still further, we might note that even the idea that stars and planets were material objects would have been foreign to ancient thought. One did not ask questions concerning the material properties (e.g., chemical composition) of stars because stars were not even thought of as ordinary material objects made of the same "earthy" stuff as the terrestrial world. The ancients did not inquire into the behavior and history of stars as material objects because they did not conceive of these celestial bodies as being made of ordinary matter that would exhibit the same patterned behavior as that displayed by the matter of our terrestrial environment. Sun, moon, stars, and planets were seen as members of an entirely different realm. The heavenly array was assumed to belong to a wholly distinct world—a world quite different from the terrestrial world of human habitation. The heavenly realm was considered to be populated not by mere material bodies but rather by spiritual powers, gods who controlled the human environment and directed the human destiny. To this matter of astral religion we turn next.

THE DEIFICATION OF CELESTIAL LUMINARIES IN ANCIENT POLYTHEISM

Because the ancient Near Eastern world is spatially, temporally, and culturally so distant from us, it is difficult for us to capture its view of the physical world, life, humanity, or deity. It is even more difficult to summarize these matters in a few pages. Thus what follows is not so much a summary, which attempts to say a little about everything, as it is a collection of fleeting glimpses of some of the important and typical ancient Near Eastern concepts and perceptions. Before looking at concepts associated with a specific culture, we should note a few ways of looking at the world of natural phenomena that were common throughout the ancient Near East.

Modern materialism, or naturalism, assumes that the natural world can be completely described in terms of its material, physical properties and its self-regulated mechanical behavior. In bold contrast to this view of nature as an impersonal, insentient, and unresponsive machine is the view that was held in the ancient Near East—namely, that the natural world is a community of sentient and responsive *persons*. The world was held to be not an impersonal thing but rather a personal *being*—not an "It," but a "Thou."[3]

The question of why some natural event occurred was answered not in terms of some impersonal material mechanism but rather in terms of conscious, purposeful, willful behavior on the part of some personal aspect of nature. Why did the river flood? Not merely as a consequence of a heavy rainfall or snow melt, but because the river willed to overflow for some purpose, either benign or malevolent.

The personification of the natural world, its objects and its phenomena, generally led to the deification of certain forces in nature or of the visible agents of these forces. The river was not merely a stream of water; the river was a god for whom the stream of water was a manifestation. The river was not simply governed or controlled by the god; the river and the god were one, and their actions were indistinguishable. The river was not the god's agent or the revelation of his power; it was a manifestation of his very being.

Ancient mythology, then, can best be seen as an expression of the prevailing understanding of the action, interaction, and counteraction of these natural forces and agents in the form of stories (story being a characteristic ancient Near Eastern genre) about the corresponding activities of the gods who were man-

3. For an excellent discussion of the ancient Near Eastern view of nature, see Henri Frankfort et al., *The Intellectual Adventure of Ancient Man* (Chicago: University of Chicago Press, 1946).

ifest in these forces or agents. These mythological stories were not mere fantasy; they were a means of expressing the human experience with nature—a world conceived of as the manifestation of numerous personal deities whose power over human life could not be denied. (There were other than natural deities, but those are not our present concern.) The cosmos and its constituent parts were given the status of deity. This identification of nature and deity formed the heart of pagan polytheism.

Thus, the existence of the cosmos was understood in terms of the existence of deities, some of which were presumed to be self-existent or self-created and others that were presumed to be the offspring of prior deities. Some ancient cultures saw the cosmos, or selected parts of it, as a divine emanation—something that emanated from and therefore partook of the divine being itself. In this case, too, the distinction between nature and deity was effectively lost; the two were presumed to be one.

What governed the behavior of natural phenomena? Why did the cosmos act as it did? Because, said the ancients, as deity, it willed to act. The cause for its behavior was attributed to its personal character. The power to govern its own behavior was presumed to be inherent in its divine nature. Natural entities were thought to have minds of their own.

Does the cosmos, according to polytheism, have value or purpose? Of course it does. The cosmos has value because deity has value. The cosmos has purpose because deity has purpose. Pagan polytheism, however, seems to have provided little comfort or cause for peace of mind. The purposes and goals of the ancient deities appear generally to have been little more than self-gratification. They were seldom credited with much concern for ethical or moral strictures in the pursuit of their selfish purposes. Humanity was often the helpless victim of divine whim or caprice.

Among the principal deities of ancient Egypt we find the sun god Rē, who is often identified with the creator god Atum. Atum gave birth to the primeval pair of deities Shu (Air) and Tefnut (Moisture), which in turn produced Geb (Earth) and Nūt (Sky). Some mythological tales represent Nūt as a cow standing over the Earth. As mother goddess, Nūt gave birth each morning to the Sun—the "calf of gold."

The Egyptian landscape provided little shade or shelter from the sun's penetrating heat. Rē was the personification of limitless solar power, the source of light, warmth, and life. Rē was also the divine king of Egypt; the Pharoahs were said to be

sons of Rē—Rē in human form, sharing the status of deity. God and king were indistinguishable, and the king was therefore to be worshiped as a god.

The moon god Thoth had considerably less importance in the Egyptian pantheon, as did the deities associated with the stars. Their function was limited primarily to calendric matters.

Aton, the god of the solar disk, was given a very special status by Pharoah Akhenaton (his name means "effective spirit of the Aton"), who reigned just prior to Tutankhamen, about a century before the Exodus. Akhenaton declared that Aton was the only god and demanded that the people worship him exclusively. Though Akhenaton's attempt at monotheism failed, it did represent a distinct break from the polytheism of the day. In a magnificent hymn, Akhenaton sings the praises of his sun-god:

> Thou appearest beautifully on the horizon of heaven,
> Thou living Aton, the beginning of life! . . .
> Thy rays encompass the lands to the limit of all that thou hast made. . . .
> How manifold it is, what thou hast made!
> They are hidden from the face (of man).
> O sole god, like whom there is no other!
> Thou didst create the world according to thy desire,
> Whilst thou wert alone:
> All men, cattle, and wild beasts,
> Whatever is on earth, going upon (its) feet,
> And what is on high, flying with its wings. . . .
> Thou makest the seasons in order to rear all that thou hast made,
> The winter to cool them,
> And the heat that they may taste thee.
> Thou hast made the distant sky in order to rise therein,
> In order to see all that thou dost make.
> The world came into being by thy hand,
> According as thou hast made them.
> When thou hast risen they live,
> When thou settest they die.[4]

While Egypt basked in the benevolence of her solar deities Rē, or Rē-Atum, or Aton, Mesopotamia stood in wary respect of nature's awesome powers—powers that sometimes gave comfort, sometimes evoked fright. Differences in geography and climate may account for the different viewpoint:

---

4. Cited by James B. Pritchard in vol. 1 of *The Ancient Near East*, (Princeton: Princeton University Press, 1958), pp. 226-30.

> The Egyptian cosmos was eminently reliable and comforting. . . . Here in Mesopotamia, Nature stays not her hand; in her full might she cuts across and overrides man's will, makes him feel to the full how slightly he matters. . . . He confronted in Nature gigantic and wilful individual powers. . . . He saw the cosmic order as an order of wills—as a state. . . . To understand nature, the many and varied phenomena around man, was thus to understand the personalities in these phenomena, to know their characters, the direction of their wills, and also the range of their powers.[5]

The four chief deities in the Mesopotamian pantheon of divinized natural powers were those representing sky (the heavens), storm, earth, and water. Anu, the sky god, was recognized as the source of all authority. Enlil, lord of the storm, symbolized the exercise of force for either good or malevolent purposes. "Mother Earth," the goddess of productivity and fertility, was known variously as Ki, Nintu, or Ninhursaga. Ea (or Enki) was the god of water and represented creativity.

The *Enuma Elish* is a Babylonian epic poem whose hero is Marduk, the city god of Babylon. The experts tell us that this composition, written toward the end of the second millennium B.C., is clearly a modified version of an original in which the hero was not Marduk but Enlil ("Lord Storm"). Incorporated into this story, which is meant to exalt Marduk as a most powerful god, is a kind of creation account—a picture of how the world was formed by the divinized forces of nature. It will be worthwhile to take a brief look at this account in order to compare it with the Bible's creation story.[6]

The story begins with a watery chaos of three personified elements: Apsu (fresh water), Tiamat (the sea), and Mummu (probably fog or mist). Apsu and Tiamat give birth to Lahmu and Lahamu, the gods representing land or silt. This clearly reflects the Mesopotamians' experience of new land being formed as the Tigris and Euphrates rivers deposited silt where they emptied into the Persian Gulf.

Lahmu and Lahamu are then said to give birth to Anshar and Kishar, gods representing the horizon, which forms the boundary between earth and sky. To Anshar and Kishar is born Anu, the sky god, who in turn bears Ea, here representing earth.

5. Frankfort, *The Intellectual Adventure of Ancient Man*, pp. 126-34.

6. See Alexander Heidel, *The Babylonian Genesis*, 2d ed. (Chicago: University of Chicago Press, 1951).

But now the serenity of this succession of divine generations is broken by violent conflict. Order and chaos engage in bloody battle. Ea (earth) kills Apsu (fresh water) and establishes her abode above his body. Marduk (probably Enlil in the earlier version of the *Enuma Elish*) is born to Ea. Tiamat (the sea) takes a new husband, Kingu, in place of the slain Apsu.

As the god of the sea, Tiamat represents malevolence and chaos. She must, therefore, be challenged and subdued. First Ea confronts Tiamat, but fails. Then Anu challenges Tiamat, but even the sky god is unsuccessful. Finally Marduk is vested with great power and authoritative word, and he faces Tiamat, slaying the sea god and cutting her body in two. With one half he forms the sky, and with the other he forms the earth. He places the celestial luminaries in the sky to establish days and months and years.

Kingu, the husband of Tiamat, is also slain, and from his blood is formed mankind, who are assigned to perform menial tasks for the gods. Far from being the crown of Creation, humanity is held to be a race of slaves for the selfish deities.

It is generally agreed that Babylonia is the birthplace of astrology. Since the time of Hammurabi (ca. 1700 B.C.), and probably even earlier, the Babylonians worshiped the stars as "gods of the night."[7] Early astrology was but part of a vast collection of omens from lunar eclipses to animal entrails that were thought to betoken either warning or encouragement. Zodiacal astrology did not appear until about the sixth century B.C. Initially applied by the Chaldeans only to affairs of the state and its rulers, zodiacal astrology (which reads great significance into the location of sun, moon, and planets within the zodiac) was later expanded by the Greeks and Romans to the affairs of all individuals.

Since it was believed that the sun, moon, planets, and stars were deities with influence over all terrestrial and human events, it became important to know their past, present, and future positions. That appears to have provided the incentive for the development and practice of mathematical astronomy. Such was the situation for nearly two millennia—from the late Babylonian period until the scientific revolution in the middle of the second millennium A.D. Even the medieval church was infected by a Christianized form of astrology, and today astrology

---

7. B. L. van der Waerden, *Science Awakening II: The Birth of Astronomy* (New York: Oxford University Press, 1974), p. 175.

is as close as your daily newspaper or your radio. The longevity of superstitious nonsense is astounding.

But even more subtle remnants of Babylonian, Greek, and Roman astrology survive. The seven days of the week bear names drawn from planetary astrology. As Cumont pointed out long ago, "When today we name the days Saturday, Sunday, Monday, we are heathen and astrologers without knowing it, since we recognize implicitly that the first belongs to Saturn, the second to the Sun, and the third to the Moon."[8] The date of December 25, chosen in the fourth century A.D. as the day to commemorate the birth of Christ, was originally the date for the annual celebration of the birth of the "new sun," the beginning of longer periods of daylight after the sun had passed the winter solstice point. Even our common vocabulary has astrological contributions. The words *lunatic, disaster, martial,* and *jovial,* for example, all have their roots in astrological lore. The first meaning of *influence* given in my dictionary is "an ethereal fluid held to flow from the stars and to affect the actions of men." And *influenza* was so named in the belief that epidemics were the result of stellar influence.

But we must return to the ancient Near East for one more contribution to the view from Palestine.

While Egypt and Mesopotamia were the major powers of the ancient Near East, Canaan (or Palestine) often served as the meeting ground for these giants—sometimes at war, sometimes engaged in friendly commerce. Canaan experienced both periods of relative independence and of domination by one external power or another. The Canaanites thus comprised a variety of people, having great diversity and fluidity of culture and religion.

In general, the Canaanites practiced the same naturalistic polytheism that their neighbors did, though with perhaps less emphasis on the identification of nature with divinity. Their chief god, El, had a status similar to that of the Mesopotamian Anu or the Egyptian Rē. El's principal offspring was Baal, representing the divinity of storm and rain (like Enlil) as well as vegetation and fertility. The chief goddesses were Asherah (mother or wife of Baal), Anath (wife-sister of Baal), and Astarte. These goddesses appear to have shared their territory as the female deities of love, sex, fertility, and war. Ritual prostitu-

---

8. Franz Cumont, *Astrology and Religion among the Greeks and Romans* (1912; rpt. New York: Dover Publications, 1960), p. 91.

tion to celebrate and encourage fertility appears to have been common practice. The Canaanites were certainly not known for a high view of morality.

The role of celestial deities in Canaanite religion is not clear from archeological evidence. Astarte, like the Mesopotamian Ishtar, was identified with the planet Venus. The name of Shapash, the sun goddess, is found in early literature, but there is little physical evidence of extensive worship of the sun or moon. Yet the worship of the heavenly hosts (the numerous celestial deities manifest in sun, moon, planets, and stars) must have been prevalent during the first millennium B.C. because of the numerous Old Testament warnings against such practices. Such astral religion would certainly be expected in the general context of the ancient Near East.

CHAPTER THREE

The Heavens according to Scripture

In an apt metaphor, John Calvin referred to Scripture as the "spectacles" that we need in order to sharpen our perception of God's revelation in human history and in the physical world. Applying that metaphor to our present concern, how can we say that Scripture clarifies our view of the heavens? What do we learn from the Bible about stars? When we view stars through the spectacles of Scripture, what do we see?

Before we can profitably seek answers to such questions, we must decide on an appropriate approach. Earnestly desiring to take the Bible seriously, we will seek to apply the general principles noted in Chapter One. But what specific program shall we employ to learn what the Bible has to say about the celestial luminaries?

We could begin by consulting a Bible concordance and preparing a list of all scriptural references to sun, moon, stars, heavens, hosts of heaven, and the like, and then develop explanatory and interpretative comments on each item. But such an approach strikes me as entirely too atomistic. It suggests breaking the Scriptures into small fragments to be interpreted in isolation from the whole, which would clearly violate their literary and covenantal integrity. Furthermore, it strikes me as a singularly dull approach.

Another method that we could employ in our quest for Scripture's teachings on stars would be to formulate certain hypotheses concerning stars and test them by searching the Bible for statements that either support or contradict them. This method also may have some merit, but it appears to me to place too much emphasis on our own hypotheses or theories, and not enough emphasis on what Scripture says, and it might present a rather strong temptation to select only those biblical statements that support our presuppositions.

The appropriate method, it seems to me, should neither pick the Scriptures apart into a list of isolated verses nor impose humanly devised themes on Scripture, forcing it to illustrate what we want it to say. Instead, I believe we should bring appropriate questions to Scripture and search for answers in a way that gives evidence of our taking the Bible seriously.

It might be objected that even the bringing of questions to Scripture gives opportunity for human bias to enter into Bible study. The selection of which questions to bring allows for the exercise of personal prejudice. Quite correct! Therefore, we must emphatically insist on qualifying these questions with the word *appropriate*. Not all questions can be profitably brought to the Bible for answers. Indeed, if we attempt to bring inappropriate questions to the Scriptures, questions that are inconsistent with the very nature of Scripture, we will risk constructing meaningless or insignificant answers at best or, more seriously, misleading, erroneous, or bizarre answers.

But how do we determine whether or not a question is appropriate? This may seem paradoxical, but I believe that in order to ask appropriate questions of Scripture we must first be thoroughly familiar with the nature of Scripture itself. Or, to put it in the terminology we used in the first chapter, before we are prepared to formulate specific questions for the disciplined academic study of Scripture, we must first have a thorough familiarity with its status, its character, and its functions. If we are to avoid prejudice in the very formulation of questions, we must already be knowledgeable concerning the principal themes of Scripture. Only in the light of such a familiarity will we be able to avoid errors in bringing questions to Scripture. Formulating appropriate questions is a vital part of taking the Bible seriously.

In formulating appropriate questions about stars, the first thing we have to take care of is the matter of terminology. There are many references in Scripture to stars, frequently in conjunction with the sun and the moon. In fact, the phrase "sun, moon, and stars" is perhaps familiar to us primarily because of the frequent biblical reference to this combination of celestial bodies. Planets do not receive specific mention; as lights in the sky they are treated simply as stars, which is quite reasonable in the historical context in which the Bible was written. For much of our discussion we need not make a distinction among sun, moon, planets, and stars. They are all celestial luminaries, and as such receive rather similar treatment in the Bible.

Speaking now of stars in this broad categorical sense of heavenly bodies, what is the principal question concerning stars to which Scripture provides an answer? I am convinced that it is the question of their *status.* As we have already noted, the concept of covenantal relationships is central to the message of Scripture. The Bible's primary function is to establish right relationships among God, the human race, and the material world. In light of this central theme of Scripture, it seems very appropriate to raise the question of the status of stars. Where do stars (or any other material object for that matter) stand in relationship to God? Where do stars stand in relationship to humanity?

The concept of the status of the material world is crucial to the remainder of this book. Not only do I judge the question of status to be an *appropriate* question to bring to Scripture; I would go so far as to say that it is the *principal* question about the material world that the Bible addresses. Closely related to the question of status are questions concerning the origin, governance, value, and purpose of the material world (here represented by stars). The remainder of this chapter will be devoted primarily to the question of status. The other questions will be treated more thoroughly in the next chapter.

SCRIPTURAL INSTRUCTION CONCERNING THE STATUS OF STARS

Stars Are Not Astral Deities

Israel lived in continuous contact with polytheistic paganism. The Egyptians, the Babylonians, and the Canaanites all had many gods—most of them manifested in the material world. These gods, though appearing in the form of familiar physical entities—rivers, seas, mountains, storms, earth, moon, sun, and stars—were presumed to be personal and powerful. They were treated as sentient, willful powers that exerted great influence on the life of humanity. To gain favor with these powers, people systematically worshiped them. Israel's pagan neighbors considered the celestial luminaries, the "heavenly array," the "hosts of heaven," to be astral deities deserving of human worship and adoration.

But Israel was called to be different, to be the elect nation of Yahweh. Repeatedly the Israelites had to be reminded of this special calling. (Their memories seem to have worked at least as selectively as ours do today.) The book of Deuteronomy presents us with an account of Moses' final instructions to the Israelites as they prepared to enter Canaan. Particularly relevant to our discussion is the following admonition:

> When you raise your eyes to heaven, when you see the sun, the moon, the stars, all the array of heaven, do not be tempted to worship them and serve them. (Deut. 4:19)

Lest we take this statement to be merely a mild suggestion rather than a stern and absolute prohibition of astral worship, note the severity of the penalty prescribed for a violation of this commandment:

> If there is anyone, man or woman, among you in any of the towns Yahweh your God is giving you, who does what is displeasing to Yahweh your God by violating his covenant, who goes and serves other gods and worships them, or the sun or the moon or any of heaven's array—a thing I have forbidden—and this person is denounced to you; if after careful inquiry it is found true and confirmed that this hateful thing has been done in Israel, you must take the man or woman guilty of this evil deed outside your city gates, and there you must stone that man or woman to death. (Deut. 17:2-5)

But, though the prohibition against idol worship and against the worship of heavenly luminaries was stern, and the penalty for violation severe, Israel repeatedly disobeyed. In 2 Kings we read of the fall of Samaria, capital of the Northern Kingdom of Israel, to the Assyrians (estimated to have occurred about 721 B.C.). The fundamental reason for this defeat is also stated: Israel failed to keep the principal covenantal requirement to worship Yahweh only.

> They despised his laws and the covenant he had made with their ancestors. . . . They . . . became empty through copying the nations around them. . . . They made themselves sacred poles, they worshiped the whole array of heaven, and they served Baal. . . . For this, Yahweh was enraged with Israel and thrust them away from him. (2 Kings 17:15-16, 18)

Thus, star worship was one of the practices that led to the fall of the Northern Kingdom of Israel.

Unfortunately, the Southern Kingdom of Judah appears to have followed suit. "Judah did not keep the commandments of Yahweh their God either, but copied the practices that Israel had introduced" (2 Kings 17:19). It is reported that during the reign of Manasseh (ca. 687-642 B.C.) many practices of Judah's polytheistic neighbors were imitated and that Manasseh "worshiped the whole array of heaven and served it. . . . He built altars to the whole array of heaven in the two courts of the Temple of Yahweh" (2 Kings 21:3, 5). For this and other covenant violations, Judah was sentenced to destruction, as the prophets had warned.

Following shortly after Manasseh, King Josiah (ca. 640-609 B.C.) sought to reverse the situation by initiating a program of covenant renewal. Under Josiah, the priest Hilkiah "did away with the spurious priests . . .who offered sacrifice to Baal, to the sun, the moon, the constellations and the whole array of heaven" (2 Kings 23:5). Josiah's successors, however, appear to have resumed the pagan practices that Josiah had attempted to abolish. The final result was as predicted: Jerusalem, capital of Judah, fell (ca. 587 B.C.) to Nebuchadnezzar, king of Babylon. The majority of the Jewish community was driven out of Judah and dispersed. Many Jews, particularly the influential leaders, were exiled to Babylon.

While in Babylon, the exiles undoubtedly became familiar with Babylonian culture, including the astral religion and the astrological practices of the Chaldeans. If, as these Babylonians believed, the stars of heaven were divine powers that influenced the affairs of humanity, then it was important, particularly in times of national anxiety, for astrologers to determine what was "in the stars."

What did the exiled Israelites think of Babylonian astrology? It is difficult to tell for certain, but there are hints that they were tempted to take it seriously and to question the power of their own God, Yahweh, who had allowed the near destruction of Israel. In bold opposition to such ideas, Second Isaiah speaks words of harsh condemnation of the pagan religion of Babylon and the sorcerers, wizards, and astrologers who serve as its priests.[1] The prophet proclaims that Yahweh alone is able to

---

1. "Second Isaiah" is the common designation of the unidentified prophet whose writings appear in chapters 40-55 of the book of Isaiah. For further information, see Bernhard W. Anderson, *Understanding the Old Testament*, 3d ed. (Englewood Cliffs, N.J.: Prentice-Hall, 1975), pp. 442-45.

influence the course of history and to redeem his people. In the face of imminent invasion by Cyrus of Persia, the Chaldeans are challenged:

> You have spent weary hours with your many advisors.
> Let them come forward now
> and save you, these who analyze the heavens,
> who study the stars
> and announce month by month
> what will happen to you next.
> Oh, they will be like wisps of straw
> and the fire will burn them. . . .
> They will all go off, each his own way,
> powerless to save you. (Isa. 47:13-15)

I will have more to say about Second Isaiah's message to Israel concerning the status of stars, but we should note at the outset that he certainly makes it clear that astrology and all astral deities are utterly powerless. The stars of the heavenly array are not gods; they have no power or influence in human affairs.

But if the sun, moon, and stars do not have the status of deity—that is, if stars are not sentient spiritual beings who have power over human affairs and who are deserving of worship—then what are they? What is the status of this heavenly array of celestial lights? For people who take the Bible seriously, the answer is obvious and can be found in many Scripture texts. We will look at several of these shortly, but for the moment let us continue to follow the history of Israel.

After the return from exile and the rebuilding of both the Temple of Yahweh and the walls of the city of Jerusalem, there was a restoration of national identity and purpose. The Jews were once again the children of Israel, the chosen people of Yahweh. In the book of Nehemiah there is recorded a magnificent ceremony of atonement. It begins by addressing Yahweh as the maker of heaven and earth.

> Yahweh, you are the only one.
> You made the heavens, the heaven of heavens, with all their array,
> the earth and all it bears,
> the seas and all they hold.
> To all these you give life
> and the array of the heavens bows down before you. (Neh. 9:6)

The remainder of the liturgy praises God for his faithful hand in the deliverance of Israel from both Egypt and Babylon. The same

God who created heaven and earth has also created a people of his own. Yahweh is equally active in natural phenomena and human history.

Back to our question: If the stars are not gods to be worshiped, what is their status? According to Scripture, stars have the status of *Creation*. Stars are a part of the created world—no more, and no less. Stars ought not to be objects of worship; indeed, they must join all other creatures in praising and worshiping the one and only Creator.

NOTE: In order to preserve an important distinction throughout this book, I have adopted the convention of capitalizing *Creation* when referring to the material product of God's creative work, while using *creation* uncapitalized when referring to the divine action that brought that product into existence. I would encourage the reader to think of *Creation* as the proper name of the physical world, the name that best indicates its status relative to God.

Stars Have the Status of Creation

The entire physical world, stars included, has the status of Creation. That is a familiar biblical theme, but because it has received such widely varied interpretations within the Christian community, I would like to spend a little time exploring Scripture concerning the concept, examining a range of biblical statements about both the Creation and the Creator. We will focus on texts containing references to stars and other celestial luminaries, but I believe that what we learn from the Bible's talk about stars as created entities applies equally well to all of Creation. Outside of the human race, all of Creation has the same status; the entire physical universe stands in the same relationship to the Creator. Humanity is of course part of the Creation, but it stands before the Creator with special additional responsibilities that go beyond the scope of this discussion.

Now, where shall we look to find Scripture's references to stars as Creation and its instructions concerning what it means for something to have the status of Creation? I strongly suspect that most Bible readers would look first at the creation narratives of Genesis 1-3. However, since these particular accounts and their varied interpretations within the Judeo-Christian community play such a prominent role in current discussions about creation (the revived creation/evolution debate, for example) we will deal with Genesis 1 in a separate chapter. In any case, the book of Genesis is not the only place in which the Bible speaks of creation. The most extensive Old Testament references are found in Job, Isaiah, and the Psalms. We will explore these first, and then consider what is contributed by the New Testament.

Job

The book of Job, which may have been written about the time of the exile, is a fascinating piece of literature. It contains elements of narrative, dialogue, poetry, tragedy, drama, and even comedy.[2] Because of its highly artistic character and its rich variety of literary elements, we must be careful in quoting isolated passages; but let us note just a couple of references in their context.

In the first cycle of the dialogue, Job's reply to Bildad's speech reveals his awe at the power and sovereignty of God. Among many illustrations are these:

> The sun, at his command, forbears to rise
> and on the stars he sets a seal.
> He and no other stretched out the skies,
> and trampled the Sea's tall waves.
> The Bear, Orion too, are of his making,
> the Pleiades and the Mansions of the South.
> His works are great, beyond all reckoning,
> his marvels, past all counting. (Job 9:7-10)

Thus, Job recognizes God as the Originator and Governor of the heavenly bodies as well as of the earth and earthly creatures. Yet later in the drama, when he needs to be humbled, Job gets a powerful reminder of these truths and of his own smallness before the Creator when the writer has Yahweh himself speaking to Job and his critics. Who can forget the power of these rhetorical questions:

> Where were you when I laid the earth's foundations?
> Tell me since you are so well informed! . . .
> Can you fasten the harness of the Pleiades,
> or untie Orion's bands?
> Can you guide the morning star season by season
> and show the Bear and its cubs which way to go?
> Have you grasped the celestial laws? (Job 38:4, 31-33a)

---

2. See Leland Ryken, *The Literature of the Bible* (Grand Rapids: Zondervan, 1974), p. 109.

Job's reply to the speeches of Yahweh is one of fitting humility and repentence. It begins, "I know that you are all-powerful: what you conceive, you can perform" (Job 43:2).

The book of Job deals with the relationship of God and mankind, raising the specific question: Why do the righteous suffer?[3] It is in the course of exploring the answer to that question (though never fully resolving it) that Job considers God's sovereign power over all of his Creation, his creatures included. Nowhere in the book of Job do we find any abstract argument to demonstrate that God is the Creator or to explain the manner of his creative activity. Rather, we are led to stand with Job in awe and respect for the Creator of the stars, of the earth, and of ourselves. The emphasis is on relationship, on status. God is the sovereign Creator; stars, earth, and humanity stand as dependent creatures; both stars and we have the status of Creation.

The references to God's creative work are highly figurative in character and serve no scientific function; rather, they are meant to answer the question of who's in charge here. Who is in ultimate control of man and beast, the earth, and even the celestial bodies? The answer: Yahweh the wise, powerful, and righteous One. The emphasis here is clearly placed on the relative status of Creator and Creation, not on the mode of creation or the mechanism of God's creative action.

Second Isaiah

Another portion of Scripture rich in references to the Creator is "The Book of the Consolation of Israel," chapters 40-55 of Isaiah.[4] Written toward the end of Israel's exile in Babylon, Second Isaiah has as one of its principal themes the praise of Yahweh, who is now redeeming his created nation from a second captivity. Israel evidently needed some reminders concerning the identity of their God, because the prophet (and skilled poet) repeatedly appears to address the question of who Yahweh is and what he is like.

Second Isaiah states the answer quite emphatically: Yahweh is the Creator and Redeemer of his people Israel. Yahweh created the nation of Israel, redeemed it once from Egypt, and is now redeeming it from exile in Babylon. What gives Yahweh the credentials to be Israel's Redeemer? The fact that he is Israel's Creator. Moreover, Yahweh is the Creator not only of Israel

3. See Anderson, *Understanding the Old Testament*, pp. 548-62.
4. See Anderson, *Understanding the Old Testament*, pp. 437-70.

but of the whole universe! Therefore, Israel, praise Yahweh the Creator of Israel; praise Yahweh the Creator of all things—even the heavens and the earth.

> Did you not know,
> had you not heard?
> Was it not told you from the beginning? . . .
> He has stretched out the heavens like a cloth,
> spread them like a tent for men to live in. (Isa. 40:21-22)
>
> Lift your eyes and look.
> Who made these stars
> if not he who drills them like an army,
> calling each one by name?
> So mighty is his power, so great his strength,
> that not one fails to answer. (Isa. 40:26)
>
> I, myself, Yahweh, made all things,
> I alone spread out the heavens.
> When I gave the earth shape, did anyone help me? (Isa. 44:24)
>
> I it was who made the earth,
> and created man who is on it.
> I it was who spread out the heavens with my hands
> and now give orders to their whole array. (Isa. 45:12)
>
> Yes, thus says Yahweh,
> creator of the heavens,
> who is God,
> who formed the earth and made it,
> who set it firm,
> created it no chaos,
> but a place to be lived in:
> "I am Yahweh, unrivaled . . . " (Isa. 45:18-19)

The relative status of God, mankind, and the material world are once again proclaimed. As we saw in the book of Job, the emphasis is not on the mode or chronology of creation, not on the structure or behavior of the Creation, but rather on the status of the Creator—his authority over the entire Creation, both in heaven and on earth, both in "natural" phenomena and in human history. And because he is the Creator, he also has the power to be Israel's Redeemer.

In beautiful and rich poetic language, Yahweh is praised for his greatness. But not only his people are called to praise God for his mighty deeds; the rest of Creation as well is called to join in the celebration of Yahweh's goodness:

> Shout for joy, you heavens, for Yahweh has been at work!
> Shout aloud, you earth below!

> Shout for joy, you mountains,
> and you, forest and all your trees!
> For Yahweh has redeemed Jacob
> and displayed his glory in Israel. (Isa. 44:23)

Psalms in Praise of the Creator

Neither stars nor any other parts of the material world have the status of deity. Stars are not gods; there is only one God: Yahweh, the Creator. Only he is to be worshiped and praised. It is his majesty and power that are displayed in the heavens and on earth.

These sentiments, and many more, are beautifully expressed in the Psalms, the lyric poetry of the worshiping community of Israel. The psalms we have in the Bible not only aid us in expressing the praise that God deserves but also remind us of the manifold reasons why we praise him. Let's look at a few examples that illustrate the theme of praise to the God who is the Creator of the heavens.

The status of Israel's Creator-God relative to the imagined gods of her polytheistic neighbors is stressed in Psalm 96:4-6:

> Yahweh is great, loud must be his praise,
> he is to be feared beyond all gods.
> Nothingness, all the gods of the nations.
> Yahweh himself made the heavens,
> in his presence are splendor and majesty,
> in his sanctuary power and beauty.

Though Israel is surrounded and even infiltrated by pagan polytheism, it must be rejected. While the temptation may seem remote to us in our culture, the attractiveness of the pagan gods was almost overwhelming to ancient Israel.

Clearly, the gods of ancient paganism are not the creators of the heavens. But the celestial luminaries, along with all other corporeal bodies in the cosmos, do need a Creator. Thus, another theme that we find throughout Scripture, including the Psalms, is the theme that everything exists only because of God's creative work, only as a product of God's will as expressed by his effective word. The universe is not self-created, but God-created. From the many passages that incorporate or imply this truth, let's consider the poetic expression of Psalm 33:6-9:

> By the word of Yahweh the heavens were made,
> their whole array by the breath of his mouth;
> he collects the ocean waters as though in a wineskin,
> he stores the deeps in cellars.

> Let the whole world fear Yahweh,
> let all who live on earth revere him!
> He spoke, and it was created;
> he commanded, and there it stood.

From this, and a multitude of similar biblical statements, we learn that the material world is neither some eternally self-existent entity alongside of God nor some self-created entity independent of God. The cosmos does not stand alongside God as an equal or in place of God as a substitute. The cosmos has the status of Creation: it stands under God in total dependency on him, its very existence dependent on divine action.

How did God bring it into existence? What is the mode of God's creative activity? Poetic references to divine speech, while sufficient to establish the essential relationship of Creator and Creation, give us no clues as to the mechanism of divine origination, preservation, or governance. But we should not be disturbed by that omission. Surely the character of our relationship to the Creator is far more important than the particular mechanism by which he chose to establish and maintain that relationship. Perhaps we would be unable to comprehend the mechanism even if we were told about it.

Israel lived in the context of a polytheism that blurred the distinction between the world of material objects and the pantheon of deities.[5] Nature and deity were thought to be essentially one. Against this background, Scripture's declaration that the cosmos has the status of Creation—quite distinct from the status of Creator—becomes particularly important. Those who are unfamiliar with the cultural setting of the Bible can easily overlook the significance of this distinction. C. S. Lewis points this out in the context of his discussion of nature as portrayed in the Psalms:

> The Jews, as we all know, believed in one God, maker of heaven and earth. Nature and God were distinct; the One had made the other; the One ruled and the other obeyed. This, I say, we all know. But for various reasons its real significance can easily escape a modern reader if his studies happen not to have led him in certain directions.[6]

---

5. For a discussion on pantheism, which also blurs this distinction, see Langdon Gilkey, *Maker of Heaven and Earth* (Garden City, N.Y.: Doubleday, 1959), pp. 58-63.

6. Lewis, *Reflections on the Psalms* (New York: Harcourt Brace Jovanovich, 1958), p. 77.

Lewis goes on to look at some of the implications of that distinction:

> To say that God created Nature, while it brings God and nature into relation, also separates them. What makes and what is made must be two, not one. Thus the doctrine of Creation in one sense empties Nature of divinity. How very hard this was to do and, still more, keep on doing, we do not now easily recognize.[7]

Is that so? Do we really still have difficulty in recognizing the distinction between nature and divinity? Is Lewis correct in suggesting that some form of polytheism or pantheism is still with us? Surely, you say, that isn't a problem in Western culture; in the East, perhaps, but not in the enlightened West! Or is it? Listen carefully to references to nature in the popular media of today. Listen and note how often "nature" is spoken of as if it were "Nature"—a sentient personal being with irresistible powers.[8] "You can't fool Mother Nature." "Nature will take its course." Perhaps Lewis is correct after all. That distinction between nature and divinity is still being carelessly blurred. Or, as we will see in a later discussion on naturalism, some blur that distinction purposely—treating nature not so much as a personal deity, as in traditional paganism, but viewing it as an impersonal, machine-like substitute for deity.

But enough digression. We noted earlier that Job expressed awe of God for his mighty acts of creation. Perhaps the most familiar expression of such awe is provided in Psalm 8, in which the psalmist describes his experience in looking up at the night sky:

> I look up at your heavens, made by your fingers,
> at the moon and stars you set in place—
> ah, what is man that you should spare a thought for him,
> the son of man that you should care for him? (Ps. 8:3-4)

If any one of us has looked into the starry heavens and failed to experience that same sense of awe, or failed to marvel that the Creator of planets, stars, and galaxies should express care and

7. Lewis, *Reflections on the Psalms,* p. 80.

8. I will adopt the convention of capitalizing *Nature* when referring to the material world conceived as an independent, autonomous, personal, or divine entity, while using *nature* uncapitalized when referring to the material world without any commitment to a particular status.

love to us on a person-to-person basis, then we have failed miserably indeed.

Psalm 136 is a litany of thanksgiving to Yahweh for his marvelous deeds in the Creation and in his loving faithfulness to his people Israel. Selecting the portion that speaks of the creation of the heavens and the earth, we read,

> His wisdom made the heavens,
> his love is everlasting! . . .
> He made the great lights,
> his love is everlasting!
> The sun to govern the day,
> his love is everlasting!
> Moon and stars to govern the night,
> his love is everlasting! (Ps. 136:5, 7-9)

Similarly, in Psalm 147 the psalmist leads us in singing a hymn to the all-powerful "Yahweh, Restorer of Jerusalem." (Verse 2 suggests that this psalm was written after the return from exile in Babylon.) How great is Yahweh?

> He decides the number of the stars
> and gives each of them a name;
> our Lord is great, all-powerful,
> of infinite understanding. (Ps. 147:4-5)

But the psalmist does not leave us with only a distant and remote Creator who ceases his activity after calling Creation into existence. This same Yahweh is also the Governor of his Creation and the Provider for his people. Poetry, with its rich imagery, provides a particularly well-suited vehicle in which to carry such high praise. Listen to the praise of God for his continuing moment-by-moment action in the familiar world of our own experience:

> He gives an order;
> his word flashes to earth:
> to spread snow like a blanket,
> to strew hoarfrost like ashes,
> to drop ice like breadcrumbs,
> and when the cold is unbearable,
> he sends his word to bring the thaw,
> and warm wind to melt the snow. (Ps. 147:15-18)

God's activity as Governor and Provider of his Creation is described in the same way as his activity as Originator and Preserver. Note this similarity also in selections from Psalm 104:

> You stretch out the heavens like a tent,
> You build your palace on the waters above. . . .
> You fixed the earth on its foundations,
> unshakable for ever and ever. . . .
> You set springs gushing in ravines,
> running down between the mountains,
> supplying water for wild animals. . . .
> From your palace you water the uplands
> until the ground has had all that your heavens have to offer. . . . (Ps. 104:2, 5, 10, 13)

God is the transcendent Creator who is distinct from his Creation, but he is not isolated from it. He is immanent in his activity as Governor and Provider. He is as dynamically present in the routine phenomena of today as he was in any spectacular act in the past. The Psalms, for example, give him equal praise for both, and describe both sorts of activity in the same poetic manner.

Among the several references in the Psalms to the heavens as God's Creation, there is yet one more consequence of this Creation status that deserves our attention. While stars and the other members of the heavenly array are neither gods nor parts of God, and while they have neither the ability to create themselves nor the power to sustain or govern themselves, they are, like us, called upon to display the wisdom and might of the Creator. Whatever the stars and sun and moon are, they are because of God's effective will. Their marvelous properties and behavior stand not as a testimony to their own achievement but to God's. Their very existence serves as a means of praise to God. Listen to the psalmist call the whole Creation to praise its Creator:

> Praise him sun and moon,
> praise him shining stars,
> praise him highest heavens,
> and waters above the heavens!
>
> Let them all praise the name of Yahweh,
> at whose command they were created;
> he has fixed them in their place forever
> by an unalterable statute. (Ps. 148:3-6)

Regarding this important point, Lewis notes that

> the same doctrine which empties Nature of her divinity also makes her an index, a symbol, a manifestation of the

> divine. . . . The doctrine of Creation leaves Nature full of manifestations which show the presence of God, and created energies which serve Him. . . . By emptying Nature of divinity—or, let us say, of divinities—you may fill her with Deity, for she is now a bearer of messages. . . . Another result of believing in Creation is to see Nature not as a mere datum, but as an achievement.[9]

But isn't that precisely what David said in the opening line of Psalm 19—

> The heavens declare the glory of God,
> the vault of heaven proclaims his handiwork.

From these familiar Old Testament themes we move on to consider what the New Testament tells us about God the Creator and the stars of his Creation. We have noted the Old Testament emphasis on the message that Yahweh is the Creator-Redeemer of Israel; as the Creator of the heavens and the earth, Yahweh has both the authority and the power to be Israel's Redeemer. In the New Testament, God as the Redeemer is fully revealed in the person and work of Jesus Christ. But what about God as Creator? What new dimension might the New Testament provide for our understanding of the relationship between the Creator and his Creation?

Creation in the New Testament

If we were to search the New Testament for extensive discussion about the manner in which God performed and is performing his creation of the starry heavens, we would be disappointed. As in the Old Testament, so in the New, the emphasis continues to be on the identity of the Creator and on his relationship to the Creation.

Does the world need a Creator? Of course it does! The apostle Paul, for example, states that the reality of God as Creator is sufficiently manifest in the Creation that we are without excuse if we fail to recognize it:

> Ever since God created the world his everlasting power and deity—however invisible—have been there for the mind to see in the things he has made. (Rom. 1:20)

But for the mind to see, it needs the eyes of faith, available as a gift from God to all who will accept it.

---

9. Lewis, *Reflections on the Psalms*, pp. 81-83.

> It is by faith that we understand that the world was created by one word from God, so that no apparent cause can account for the things we can see. (Heb 11:3)

Here the writer of Hebrews is, I believe, reminding us that the existence of the visible universe is not self-explanatory. There are no apparent, proximate, or "natural" causes for the existence of the material world. We can see that the cosmos exists, but there is nothing within the universe itself that explains why; to identify the cause for its existence requires an act of faith. True faith allows us to see the true Creator throughout all of Creation.

And who is the Creator? The same one who is Redeemer! And just as Christ is now revealed as the perfecter of redemption, so also he is revealed as the effecter of creation. A theme we noted in Second Isaiah appears repeatedly in the New Testament: the credentials—the authority and power—of the Savior are established by identifying him as the Creator. A few examples will help to illustrate this.

Psalm 102 ends with an appeal to Yahweh as the Creator of heaven and earth. The writer to the Hebrews quotes this passage in order to establish the identity of the Son, applying the passage to Christ rather than to Yahweh:

> It is you, Lord, who laid earth's foundations in the beginning, the heavens are the work of your hands; all will vanish, though you remain, all wear out like a garment; you will roll them up like a cloak, and like a garment they will be changed. But yourself, you never change and your years are unending. (Heb. 1:10-12)

The prologue to the Gospel of John, which speaks so eloquently of the Word that existed from eternity and became flesh in Jesus Christ, also speaks of his role as Creator.

> Through him all things came to be,
> not one thing had its being but through him. (John 1:3)

This is very similar to the wording of Hebrews 1:2, which speaks of God's Son, "the Son . . . through whom he made everything there is."

In his letter to the Colossians, Paul also identifies Christ as the Creator:

> He is the image of the unseen God
> and the first-born of creation,
> for in him were created
> all things in heaven and on earth:

> everything visible and everything invisible,
> Thrones, Dominations, Sovereignties, Powers—
> all things were created through him and for him.
> Before anything was created, he existed,
> and he holds all things in unity. (Col. 1:15-17)

As the Old Testament proclaimed, so the New Testament reveals that Creator and Redeemer are one. The relationship between the Creator and his Creation is fulfilled in the relationship between the Redeemer and his redeemed. Thus the end of creation is the expression of love that culminates in the redeeming work of Christ.

Sun, moon, and stars are part of the Creation redeemed by Christ. They are not celestial powers or astral deities in competition with the Savior. They are more like servants, united with all Creation to realize God's purposes.

CHAPTER FOUR

The Form and Content of Scriptural Talk about Creation

By reviewing passages from Job, Second Isaiah, the Psalms, and the New Testament, we have been reminded that the Bible makes it clear that God, now fully revealed in Christ, can be our Redeemer because he is first our Creator. God is the Creator, and the entire cosmos and all of its creatures are his Creation—which is to say that the material world in its entirety is dependent on God for its existence, its governance, its value, and its purpose. As God's creatures, we are privileged to experience the richness of that relationship. But let us pause, now, to reflect a bit further on both the form and content of the Bible's talk about God's creative activity and the Creation that testifies to it.

MATTERS OF FORM

The reality of divine action in the material world is never questioned in the Bible; on the contrary, God's control of all material forces is a frequently recurring theme in Scripture. But observe the manner or form in which this divine action is most commonly described:

> You visit the earth and water it,
>   you load it with riches;
> God's rivers brim with water
>   to provide their grain.
>
>   This is how you provide it:
> by drenching its furrows, by leveling its ridges,

> by softening it with showers, by blessing the first fruits. (Ps. 65:9-10)

In poetic form and richly figurative language, the familiar phenomena of rivers, rain, and fields of growing grain are presented as evidence that God is at work in our world. God's governance of his Creation is plainly visible to all who have the eyes of faith to see it. In a passage reminiscent of Psalm 135:7, the prophet Jeremiah expresses the reality of divine activity governing the familiar everyday phenomena of meteorological events:

> When he thunders
> there is a tumult of water in the heavens;
> he raises clouds from the boundaries of the earth,
> makes the lightning flash for the downpour,
> and brings the wind from his storehouse. (Jer. 51:16)

We see, then, that the Creator's work as Governor of his Creation is most frequently portrayed in the form of poetry in which God's actions are presented in highly figurative and anthropomorphic language. But what about his work as Originator? In what form and language does Scripture tell us of God's calling his Creation into existence? From the previous chapter we already know the answer: in the same figurative, anthropomorphic, and poetic way. In addition to the passages already quoted, many others could be cited to illustrate Scripture's customary manner of describing God's creative activity. Second Isaiah, for example, speaks of the Creator who has "stretched out the heavens like a cloth, spread them like a tent" (40:22) and who "created the heavens and spread them out" (42:5). Speaking for Yahweh, the prophet says,

> I it was who made the earth
> and created man who is on it.
> I it was who spread out the heavens with my hands,
> and now give orders to their whole array. (Isa. 45:12)

The psalmists also use a number of anthropomorphic figures of speech in referring to the Creator's work. Psalm 8:3 expresses wonder concerning the heavens "made by your fingers." And in Psalm 102:25 we read,

> Aeons ago, you laid the earth's foundations,
> the heavens are the work of your hands.

Similarly, in Psalm 104:

> You stretched out the heavens like a tent. . . .
> You fixed the earth on its foundations.

Elsewhere in the Psalms, a different figure of speech is used. Rather than speaking of God creating through the use of his "fingers" or "hands," the psalmist portrays him as creating by verbal fiat, by the spoken command:

> By the word of Yahweh the heavens were made,
> their whole array by the breath of his mouth. . . .
>
> He spoke and it was created;
> he commanded and there it stood. (Ps. 33:6, 9)

Still more figures of speech are used in Scripture. In addition to "fingers," "hands," and "word," we find "wisdom," "discernment," "knowledge," and "power."

> By power he made the earth,
> by his wisdom he set the world firm,
> by his discernment spread out the heavens. (Jer. 51:15)
>
> By wisdom, Yahweh set the earth on its foundations,
> by discernment, he fixed the heavens firm.
> Through his knowledge the depths were carved out,
> and the clouds rain down the dew. (Prov. 3:19-20)

In typically ancient Hebrew fashion, the Bible speaks of the reality of divine action in the origination, preservation, and governance of the material world in the form of verbal pictures. God "sets the earth on its foundations." He "spreads out the heavens with his hands." The heavens were created "by the word of Yahweh." Through God's knowledge "the depths were carved out."

The message that God is the Creator and that the cosmos is his Creation is conveyed by the vehicle of artistic literature written in the language of figurative and picturesque speech, not by the vehicle of intellectual discourse written in the technical language and style that we members of twentieth-century Western culture have come to expect. Our heritage of abstract analytical discourse is rooted in the thought patterns of ancient Greece, not of ancient Palestine.[1] Once we come to appreciate that difference, we will be better prepared to formulate appropriate questions to bring to Scripture for answers.

The pictorial, anthropomorphic, figurative language of the Bible is an excellent vehicle for conveying answers to questions

---

1. To gain an appreciation of this difference, see Nahum Sarna's *Understanding Genesis* (New York: Schocken Books, 1970), which provides an insightful commentary on Genesis from a Jewish perspective.

about the character of the relationship of God to his Creation but is singularly unsuitable for the purpose of conveying answers to technical questions about the means by which God has established and continues to maintain that relationship. If, for example, we ask whether the material world is an independent autonomous machine driven solely by purposeless internal material powers, we will find an answer. If, on the other hand, we ask what specific mechanism God uses to govern the formation of clouds and the various forms of precipitation, we will find the Bible silent. We must respect that silence. If we fail to do so, if we force the Bible to speak to such a technical question, we will end up with the distorted and bizarre concept of God literally having bins filled with wind, snow, and hail from which he occasionally releases the desired amounts and dumps them on the earth below (see Job 38:22 and Pss. 135:7 and 147:15-18). Bringing appropriate questions to the Bible leads to a harvest of beautiful and powerful answers; inappropriate questions are the seeds of nonsense.

If God's creative action is portrayed in pictorial, anthropomorphic, and figurative language, what about the Bible's portrayal of the Creation? In what form or manner does the Bible speak about the material world itself? Quite obviously, the Bible does not describe the Creation in the language of modern natural science. But neither does the Bible speak in the language of ancient science. We don't find anything like the elaborate mathematical computations of ancient Babylonian astronomy in the Bible. The Bible is ascientific; it expresses no interest in either ancient or modern science. It does not speak unscientifically in the sense of speaking in ways that modern science would judge to be mistaken; nor does it speak antiscientifically in the sense of rejecting what empirical study would indicate. Rather, it speaks nonscientifically, or ascientifically. It views Creation not from a theoretical or analytical vantage point but from a common experiential perspective. Biblical speech about the nature of the Creation is understandable in all ages because it comes in the language of ordinary human experience.

Why is this the case? Why is scientific language avoided in the Bible? I can only guess at the answer, but let me propose some possibilities. If the Bible had been written in the language of the science of its day, it would have made sense to only a small minority of people, then or now. On the other hand, the language of common experience is spoken by all. Furthermore, if the Bible had been written in the language of the science of its day, then it might now be obsolete or incorrect, misleading its readers. However, I think there is a still more important and fundamental reason. Natural science, particularly the twentieth-century Western variety, treats the material world without regard to any nonmaterial influence or relationship, as a self-contained entity. Scientific methodology is incapable of investigating the nature of external, nonmaterial influences, and so it is intentionally limited to the study and description of internal relationships, properties, and patterns of behavior. Scripture, on the other hand, emphasizes the matter of relationships to God. The Bible shows no interest in the material world in isolation from God; in fact, it shows interest in the material world only because of its relationship to God, its Creator—only because it has the status of Creation. Thus the language of natural science is quite useless for the principal purposes of Scripture.

I will have more to say about natural science later, but for now let's return to matters of Scripture—particularly the content of its references to the Creation and to the creative activity of God.

MATTERS OF CONTENT

The Character of God's Creative Activity

We have noted several times already that Scripture plainly teaches that God is the Creator of all things. We now ask what it is that the Creator does. What divine activities are performed by the Creator? What is the scope or breadth of God's creative work? In a sense we have already covered the several parts that together constitute the biblical answer to this question, but let us now gather them together in order that we may see the full picture.

What does the Creator do? Many different things. The Creator's activity is multidimensional, not merely of one kind. To know what it means for God to be the Creator is to be familiar with all of these aspects or dimensions. Overlooking any one of them leaves us with an incomplete view; overemphasizing any one component produces a lopsided view.

Extensive reading from Scripture leads me to identify four categories of God's work as Creator. If asked to state what the Bible teaches us about the acts of the Creator, I suppose that most of us would begin our response by saying that as Creator, God called the whole world—"the heavens and the earth," as the Hebrews put it—into being. God is the Originator of all things; it was he who called heaven and earth into existence *ex nihilo,* from nothing. And such an answer would be correct—not complete, but correct as far as it goes. The Creator revealed by Scripture is indeed the Originator of the whole Creation. Recall, for example, these familiar words:

> It is by faith that we understand that the world was created by one word from God, so that no apparent cause can account for the things we can see. (Heb. 11:3)

Or remember these words from the prologue to the Gospel of John:

> In the beginning was the Word. . . .
> Through him all things came to be,
> not one thing had its being but through him. (John 1:1, 3)

To be the Creator, then, is to be the Originator of the cosmos, the One who called it into being out of nothing. But there is more. The Creator revealed by Scripture is also the Preserver or Sustainer of his Creation. Once called into being, the cosmos is still dependent upon divine action for its continued existence. If this were not so, we would then have a material world existing independently of God, and he would no longer be sovereign over it. But the Creator of which the Bible speaks is neither temporally nor spatially remote; on the contrary, he acted not only in an instant of origination but at every moment: he is continually active in sustaining and preserving the very existence of that which he originated. As the apostle Paul stated to the Council of the Areopagus in Athens,

> Since the God who made the world and everything in it is himself Lord of heaven and earth, he does not make his home in shrines made by human hands. . . . In fact he is not far from any of us, since it is in him that we live, and move, and exist. (Acts 17:24, 28)

And in the letter to the Hebrews, we are told of the fullness of God's revelation in the Son,

> through whom he made everything there is. He is the radiant light of God's glory and the perfect copy of his nature, sustaining the universe by his powerful command. (Heb. 1:3)

Throughout the Bible we are presented with the picture of a Creation that depends not only on God's past act as Originator but also on his continuing action as its Preserver. Thus, I believe, if God were to cut off his relationship to the Creation—if he were to cease being the Creator—we, and indeed all of Creation, would cease to be anything. If God as Creator is the Preserver and Sustainer of our very being, then the whole cosmos is as dependent on God's present action of sustaining its existence as it is on God's past action of originating its existence *ex nihilo*. If God were not at this very moment acting in the capacity of Creator, we would not at this very moment even exist.

But there is still more that the Creator does. The Creator is also the Governor of his Creation's behavior. As is the case for his work of origination and preservation, so also God's governance of the behavior of the material world has been expressed in the pictorial, anthropomorphic language so frequently found in Scripture. And while this vehicle of poetic literature and this packaging of anthropomorphic pictures does not serve well to carry technical information on the "mechanism" of divine governance, it does serve exceptionally well to convey the idea of relationship. God governs; the cosmos is governed. God directs; the cosmos follows. God commands; the cosmos obeys. If you wish to sense the scope of that relationship, read chapters 38 and 39 of the book of Job, or read Psalm 104 from beginning to end. Can there remain any doubt that Scripture teaches that the Creator acts not only as Originator and Sustainer but also as Governor, directing the manner of behavior for wind, lightning, streams, plants, animals, fish, sun, moon and stars—everything in heaven and earth? Recall how the words of Psalm 147 speak so directly of God's action as Governor of his Creation:

> He gives an order;
> his word flashes to earth;
> to spread snow like a blanket,
> to strew hoarfrost like ashes,
>
> to drop ice like breadcrumbs,
> and when the cold is unbearable,
> he sends his word to bring the thaw
> and warm wind to melt the snow. (Ps. 147:15-18)

Such passages offer no contribution to the scientific understanding of the mechanism of precipitation, but neither do they provide any discouragement or hindrance to empirical investigation. Biblical poetry is the vehicle of God's message concerning the relationship of the material world to its Creator; it reveals nothing about the properties or behavior of matter in itself or of the quantitative internal relationships between material properties and behavior or of proximate cause-effect relationships within the material world. The Scriptures speak not of internal relationships within the material world alone but of the ultimate *external* relationship—the relationship between the Creation and its Creator. And precisely because natural science has no way of handling the dimension of divine direction of material behavior, it would appear to me that natural science offers relatively little aid in understanding scriptural references to God's creative activity. Natural science excludes from its consideration the very dimension that the Bible emphasizes. As we will note later, this doesn't make either scientific description or biblical proclamation incorrect; it merely exposes the incompleteness of each.

To complete our list of the categories of divine activity carried out by God as the Creator, we add the category of providence. As Creator, God not only originates, preserves, and governs his Creation but also provides for its needs. This too is part of the covenant relationship that God has established with his creatures; he has promised, within the context of his sovereign purposes, to provide for the needs of his creatures in such a way that the end, or goal, of creation will be accomplished.

God's providence for the needs of his creatures takes many forms. He supplies strength of character to face temptations, courage to undertake a difficult task, comfort in time of grief, food for the nourishment of both man and beast. For the faithful provision of all of these needs we return thanks—even for those things that appear to be part of the ordinary (or "natural") sequence of events in the material world.

> Sing to Yahweh in gratitude,
> play the lyre for our God:
>
> who covers the heavens with clouds,
> to provide the earth with rain,
> to produce fresh grass on the hillsides
> and the plants that are needed by man,
> who gives food to the cattle
> and to the young ravens when they cry. (Ps. 147:7-9)
>
> All creatures depend on you
> to feed them throughout the year;
> you provide the food they eat,
> with generous hand you satisfy their hunger. (Ps. 104:27-28)

Once again we see that the Bible speaks directly to the matter of the relationship of the Creator to his creatures as manifested in the scope of divine action performed on their behalf. Though the means or material mechanism (patterned behavior of matter, or physical process) is not the subject of biblical discourse in the technical language of the sciences, the reality of divine action is proclaimed.

Now let's put these four categories of God's creative activity together to form the complete picture. God is the Creator; as Creator, God is the Originator, Preserver, Governor, and Provider of the Creation. God's multidimensional activity as Creator, therefore, is not confined to some instant of "exnihilation" in either the recent or remote past. God's activity as Creator is continuously required—past, present, and future. God's creative activity is just as necessary for our daily life in the present as it was at any time in the past. Each moment of each day, for example, we experience his action as our Preserver, Governor, and Provider. This experience provides a reservoir of illustrations of God's covenant faithfulness in the same way that the experiences of the biblical historians, prophets, and poets provided the occasions for their witness and testimony to God's faithful acts in history and in the lives of his people of centuries past.

We must recognize, however, that the action of origination stands in a class of its own. Such acts are of necessity beyond our personal experience. We can at best only vaguely imagine what they are like. We have no experience that can serve as a basis for comparison, and so we should abandon any attempt to describe God's originating action in literal terms or in the technical language of the natural sciences, which depend on such experiential observation. Instead, like the writers of the Scriptures, we must resort to a less precise but vastly more powerful language—the language of artistic literature. As it is found in the Bible, artistic literature creatively and effectively employs such genres as poetry, parable, allegory, story of origins, and primal history;[2] it freely uses the picturesque speech of metaphor and pictorial or anthropomorphic representations of divine behavior; on occasion it even borrows imagery from extrabiblical

---

2. The terms "primal history" and "primeval history" are commonly used to identify the genre of Genesis 1-11. We will discuss this term more thoroughly in Chapter Five.

mythological literature, with which many biblical writers were familiar.[3]

The natural sciences will be of little or no help in the effort to understand origination. Events of true *ex nihilo* creation are as unknown in scientific observation as they are in ordinary human experience. As the writer to the Hebrews has reminded us, it is only by faith that we can even begin to comprehend what divine origination encompasses. We will return to a consideration of that comprehension as we explore the meaning of the Genesis 1 creation narrative in Chapter Five.

However, before we leave this discussion of the multiple categories of divine creative activity, we should make one additional observation from the viewpoint of our experience. I observed earlier that when asked to consider what God does as Creator we tend to overemphasize the aspect of origination. Perhaps that lopsided emphasis could be reduced if we simply rephrased the question and asked what it means to be God's creatures. It seems to me that the comfort derived from knowing that we are God's creatures, that we are part of God's Creation, is a comfort that draws deeply on the assurance of God's providential care, his sovereign governance, and his faithful preservation of Creation. From our daily experience of these aspects of God's creative work we will be led to ask how God can be our Sustainer, Governor, and Provider. Scripture tells us that God can be these things because he is the Originator of all Creation. Only the Originator has the credentials—the power and authority—to be the world's Sustainer, Governor, and Provider.

We have a tendency to summarize the biblical doctrine of creation by saying "The cosmos *was created* by God." That is altogether correct, of course, but the past-tense verb indicates our inclination to think of creation solely in terms of the act of exnihilation. It strikes me that the statement "The cosmos *is* God's Creation" constitutes a better summary of biblical teaching about creation. It includes the idea of *ex nihilo* origination, but goes beyond that, lending itself to a balanced consideration of the multiple categories of divine creativity. It has the form of a statement that clearly identifies the status of the material world—a status that in essence entails a relationship of dynamic dependency. To this matter of status and its consequences we now direct our attention.

---

3. See Bernhard W. Anderson, *Creation versus Chaos* (New York: Association Press, 1967).

The Status of Creation and Its Consequences

In Chapter Three we considered which sorts of questions about the material world could appropriately be brought to the Bible for answers. We noted that the most appropriate question is the question of status. Where, we ask, does the material world stand relative to deity and humanity, and Scripture answers, the cosmos has the status of Creation. Contrary to ancient polytheism, the material world does not stand among the deities as a member of the divine pantheon. Contrary to pantheism, material systems are not themselves parts of God or emanations of God. Rather, the cosmos is the product of the multidimensional action of the Creator. As Creation, the cosmos stands under the sovereign Creator as his covenant-bound servant. The status of Creation has certain consequences and implications that deserve special mention—consequences in the categories of origin, governance, value, and purpose.

That the cosmos has the status of Creation entails first of all that it is totally dependent on God for its existence, both for its coming into being and for its continuing existence. For this reason, we said, the Creator must act as both the Originator and the Sustainer of the universe. If the Creator ceased to act as Sustainer, the universe would cease to exist. Thus, the Creation has the property of being "radically contingent"—its very existence (not merely its form) requires a continuous act by the Creator. Contrary to the assumptions of various forms of atheism and materialism, the cosmos is neither self-generating nor self-sustaining.

Though the existence of the Creation is radically contingent upon God's action of preservation, we must still ascribe to that existence the qualities of objective reality and continuity of being. By "objective reality" I mean that it has an existence independent of human perception. The cosmos is not merely the product of human imagination. It has an objectively real existence distinct from God and independent of human perception. By "continuity of existence" I mean that the existence of the cosmos is continuously preserved or sustained rather than recreated anew at each moment. The concept of re-creation introduces such a discontinuity of existence that creatures could no longer be held accountable for their behavior, and God would be directly responsible for each creaturely act—even acts of disobedience. A world characterized by this lack of continuity or accountability seems clearly in contradiction to the idea of covenant responsibility that permeates the whole of Scripture.

The status of Creation also has implications concerning the governance of the material universe. By what power does the material world behave as it does? What, for example, causes the earth to move along an elliptical path as it orbits the sun? If we were to say that "gravity" is the cause of this behavior, is that really providing an answer, or merely giving a name to the answer? These are not trivial questions. For the purposes of the present discussion, let us restrict our attention to just two subsidiary issues: first, the question of whether the power that governs the behavior of matter resides within matter itself or lies outside of the material world, and second, the question of what qualities of corporeal behavior we might expect to follow from the identity and character of this governing power.

Internal or external? Where is the power that governs cosmic behavior to be found? If we take this question to Scripture, a very direct answer is readily found. As we have already noted, God is the Governor of the Creation. The cosmos behaves in patterns caused by the effective will of the Creator. Contrary to the assumptions of naturalism (or even of deism, for that matter) the universe does not cause, or regulate, or govern its own behavior by internally resident powers; rather, the behavior of the cosmos is governed by the Creator. We have already considered several passages from the Bible that leave no question on this point. The cosmos is not self-governed but God-governed. This raises many difficult and fascinating questions, but let's postpone those for later discussion.

If, as Scripture teaches, the Creation is governed by the Creator, what qualities of behavior might we expect? Before answering that question, we should note two things. First, such a question is not given high priority in biblical materials; the fact that God is Governor is clearly expressed, but his method of governing receives less attention. Second, our list of expected qualities of material behavior will unavoidably be influenced by our twentieth-century, Western, scientifically oriented culture and training. Nevertheless, I venture to claim that Scripture either warrants or readily permits the following expectations:

1. The material world will behave in an orderly, patterned manner under divine guidance.
2. The patterned behavior of matter and material systems is coherent (i.e., it is internally consistent and in accordance with patterns that are not capriciously or discontinuously changed).

3. The coherent, patterned behavior of matter and material systems is intelligible to human observers (i.e., it exhibits empirically accessible relationships that the human mind is capable of comprehending, at least in part).
4. The intelligible, coherent, patterned behavior of matter and material systems will exhibit proximate causality (i.e., there will be a systematic relationship between the properties of a material system and the consequent behavior of that system in a particular environment).

We will return to the matter of material behavior defined by these expectations in Chapter Six, where we will explore what it means to take the material world seriously as a natural scientist. Meanwhile, let's continue to look at the consequences of the material world having the status of Creation.

As Creation, the material world has value by virtue of its relationship to the Creator. If there were no Creator, as philosophical materialism would have us believe, then what would be the value of the material world? To whom would its existence, behavior, or history be of any interest or significance? Only to parts of the material world itself—those parts that we might call intelligent living beings. (I was about to call them "creatures," but of course if there were no Creator, there could be no creatures.) Thus, in the absence of a Creator, if the cosmos is to have any value at all, it must have value resident within itself alone: it must be self-valued.

While such a concept of intrinsic worth in the absence of a Creator may be satisfying to some people, it seems like a singularly empty concept to me. Carl Sagan, the Cornell University astronomer for whom the cosmos is "all that is or ever was or ever will be,"[4] appears to be content to marvel at the organizational complexity of the system of atoms and molecules of which he is made.

> I am a collection of water, calcium and organic molecules called Carl Sagan. You are a collection of almost identical molecules with a different collective label. But is that all? Is there nothing in here but molecules? Some people find this idea somehow demeaning to human dignity. For myself, I find it elevating that our universe permits the evolution of molecular machines as intricate and subtle as we.[5]

---

4. Carl Sagan, *Cosmos* (New York: Random House, 1980), p. 4.
5. Sagan, quoted in *Time*, 20 October 1980, p. 68.

In response to Sagan I would say, Yes indeed our bodies are marvelous molecular machines, and the more that we learn of their structure, their biophysical functions, and their temporal development, the more marvelous they appear. But as persons, even as "creatures," we are more than mere molecular machines. Such structural wonders as living organisms are of negligible value unless they can rightly be called creatures of the Creator. Even the human body is of infinitesimal significance unless it represents that which has been "fearfully and wonderfully made" by the Creator, unless it embodies a person capable of knowing God as Creator and Redeemer. Value, it seems to me, lies principally, perhaps even exclusively, in the dimension of relationship to God. Only in the knowledge of that relationship as revealed by Scripture do we begin to know the value of the material world or even the value of our own existence.

If the positive value or "goodness" of the universe lies in its relationship to God the Creator-Redeemer rather than in matter itself, perhaps we should make a similar statement concerning negative value, or evil. As Creation, the material world cannot be inherently evil. Contrary to the presuppositions on which various forms of dualism are based, the material world is not the manifestation of an independent, eternal power of evil in constant struggle against the power of good, which is represented by the spiritual world. Since the Creation is the product of God's action, it cannot be inherently evil, though it is obvious that the disobedience of divine imperatives is a possibility in the created world. Though every atom and molecule that makes up the human body necessarily obeys the divine law for its material behavior, the person embodied in that material system is nevertheless capable of violating his or her covenant obligation to live in obedience before the Creator. The phenomenon of human sin is universal (or at least global); the Garden of Eden narrative makes that quite clear. Thus, as goodness, or positive value, is to be found in a right relationship to God, so also evil, or negative value, is to be found in a broken relationship to the Creator. The heart of the gospel message of Scripture is that this broken relationship can be restored because God is not only Creator but in Christ is also the Redeemer.

To the list of the consequences of Creation status, let us add just one more—purpose. It would appear to me quite appropriate to ask of the Bible what the purpose of the Creation is. What purposes are being served by the material world, by matter itself with the particular properties and behavior which it exhibits? Toward what end or goal is cosmic history moving? I do not pretend to be able to provide full answers to these very important questions. My goal for the moment is simply to establish that such questions can legitimately be addressed to Scripture.

The Bible's primary function is to establish right relationships among God, humanity, and the rest of Creation. If God is purposeful (and Scripture certainly warrants that belief) and if the material world is his Creation, then we have every right to expect that his purposes will be revealed both in the Scriptures and in the world itself. In ascribing to the material world the status of Creation, the Bible presents the whole universe as a means by which God is accomplishing his plan to be known by his image-bearing creatures as the Creator-Redeemer. The end of creation is to manifest God's love, expressed most fully in Christ—his own word made flesh. Even the Creator was willing to assume material form in order to demonstrate his purposes.

In the context of perceiving the material world as the Creation which is serving the purposes of the Creator, it would seem quite reasonable to expect that the very properties, behavior, and history of the cosmos would display elements of purpose, or direction, or movement toward a destination. Therefore, to the four qualities I have already suggested we might reasonably attribute to the material world as God's Creation, I would add a fifth:

5. The intelligible, coherent, patterned behavior of matter and material systems will exhibit not only proximate causality but also purposeful, directed movement toward an ultimate destination.

Are there appropriate questions about the material world that we may bring to the Bible? Indeed there are. Not only appropriate but exceedingly significant questions—of status, origin, governance, value, and purpose—may be addressed to Scripture with the fully warranted expectation of receiving answers rich with meaning. But what about the other case? Are there also questions that are inappropriate? If so, what kinds of questions? What categories of questions about the material world ought never to be addressed to Scripture?

MATTERS ON WHICH SCRIPTURE IS SILENT

On matters of the relationship of the material world to God, its Creator, the Bible speaks both eloquently and frequently. This follows from the covenantal character and function of Scripture and ought to come as no surprise to a person who wishes to take the Bible seriously.

At the same time, however, there are other matters concerning the physical world on which the Bible is silent, or on which it speaks only incidentally or infrequently. For example, while the Bible does answer the crucial questions concerning the status and origin of the material world, it gives essentially no information about the specific physical properties of matter or material systems. And, while the Scriptures lead us to an understanding of the divine governance of Creation, they provide no technical information concerning the particular patterns of behavior that matter will exhibit. Finally, while the Bible does speak to the important questions of the value and purpose of the cosmos, it should not be expected to provide us with a universal history or a cosmic chronology outside of human experience.

Let us try to illustrate these statements with specific examples. We have already spoken at length about stars, so let us use these celestial luminaries as examples of material systems—things made out of matter—that possess certain physical properties, exhibit particular patterns of behavior, and undergo a definite sequence of changes over an extended period of time (we will take a look at the "life history" of stars in Chapter Eight).

According to the Bible, stars are not to be worshiped as deities; they have the status of Creation and are thus dependent on the Creator's sustaining action for their continuing existence. But suppose we came to the Bible with questions such as these: How large is a star? How far away from us are the stars? What is the chemical composition of a star? What is the temperature of a star's surface? We all recognize that such questions cannot appropriately be addressed to Scripture. The Bible was (purposely, I believe) never intended to be a source of answers to questions like these. If such questions are of interest to us, we will have to look elsewhere for answers—not to the Scriptures but to the Creation itself.

The Bible tells us that stars are governed by God's power according to the patterns (or "laws," if you prefer) he as the Creator has chosen. But questions about these patterns—questions, say, about the physical process by which a star's surface emits light, about changes occurring in the chemical and structural properties of stars, about what physical process is occurring within stars to generate the vast amounts of energy they radiate as light—are, like questions about the physical and spatial properties of stars, wholly inappropriate questions to bring to the Bible for answers. While the identity of the Governor is made clear, the particular patterns of divine governance are not the subject of biblical discussion. Once again, if we have a curiosity about such phenomena we must explore and observe and measure the Creation itself in order to satisfy our inquisitive minds; the Bible is not an answer book for such questions.

Sun, moon, and stars serve a number of very important functions in our lives. They provide the heat and light required for life itself and play useful roles in our lives as timekeepers and navigational aids. In addition to the practical values and purposes that members of the heavenly array serve, they share also in the value and purpose of the whole Creation as it functions throughout a purposeful, directed history unfolding as a revelation of God's plan. The Bible acknowledges the practical benefits that we experience and then leads us further to recognize that all of cosmic structure, behavior, and history is directed by God toward the fulfillment of his purposes. From the teaching of Scripture in its totality, I believe we are warranted in holding that cosmic history—the coherently patterned fabric of events throughout the spatially and temporally vast universe—is not merely the cumulative result of random accidents of nature but is rather the product of God's directing his Creation toward its end, its divinely established destination.

However, while the Bible does lead us to a comprehension of the purpose and divine-directedness of cosmic history, questions about the specific details of the historical sequence, particularly about that portion of cosmic history that lies outside the brief span of human experience, fall into the same category as questions about general physical processes and specific physical properties. If we are interested in finding the answers to questions concerning such things as the life history of stars—how they're formed, how they come to the end of their luminous life, how long they last, how old the sun and other stars are, and what role the sun's history has played in the history of the earth—we will have to turn to an observational study of Creation itself rather than to the Bible.

We do not err by asking these questions—they arise from perfectly normal human curiosity—but we commit a colossal blunder if we bring them to Scripture for answers. In the light of the time and manner in which the Bible was written, the limited vocabulary and cosmological concepts available to the writers, and the principal functions and purposes for which the Scriptures were written, I am convinced that questions about cosmic chronology and the physical history of the universe prior to human history should never be addressed to the Bible. Taking the Bible seriously requires that we bring only appropriate questions for Scripture to answer. To fail in that requirement is to invite confusion and befuddlement—visitors that ought never to be welcomed into the Christian mind. Even Christian hospitality can be carried too far!

CHAPTER FIVE

He Made the Stars Also

During the past decade I have taught more than two thousand students in an astronomy course called "Planets, Stars, and Galaxies." Toward the end of that course we study what astronomers have learned about the life history of stars, or, as the topic is more generally known, "stellar evolution." As we shall see in Chapter Eight, this topic necessarily leads to the consideration of a time scale involving billions of years. There is abundant evidence that stars have been forming during most of the multi-billion-year span of cosmic history. Some stars were formed more than ten billion years ago; others are forming at this very moment.

Though the majority of my students arrive familiar with the cosmic time scale and have no problems with it, there are always several in each class who have never given the matter serious consideration (assuming it, perhaps, to be mere nonsense) or who, having considered it, have very consciously rejected it in the sincere Christian belief that it is clearly contradicted by the teaching of the Bible, particularly by Genesis 1, which says simply, "He made the stars also." It is especially for these students, and for the large number of Christian believers that they represent, that I am writing this chapter.

My own background and training have, I believe, provided me with an empathetic understanding of the struggle that many Christians experience in their efforts to integrate their Christian faith in the Creator with their scientific study of the Creation. From early childhood I was taught to believe in God as the

Creator of heaven and earth as expressed, for example, in Genesis 1. But more specifically, I was taught that the correct way to understand Genesis 1 is to treat it as a chronicle of discrete creative acts of God—a list of historical events that occurred in precisely the order and time frame given there. Creation, I was told, must be understood as a series of extraordinary acts that took place within a six-day period, after which God stopped creating.[1] I was taught to interpret Genesis 1 in a literalistic manner—to treat it as a verbal photograph that records, in a very concise manner, exactly what happened during a most unusual week just a few thousand years ago.

I have since come to the conclusion that such an interpretation is naive—though in using that term I certainly do not wish to imply a lack of intelligence on the part of my instructors; nothing could be farther from the truth. (Intellectually brilliant people can be just as naive as we ordinary people can.) Rather, I use the word *naive* to connote an innocent lack of awareness of certain auxiliary information that must be incorporated into the process of biblical interpretation. Such naivete can yield to enlightenment, however, provided that one has a genuine desire to take the Scriptures seriously—even if that entails the relinquishing of a favorite tradition-laden interpretation in order to increase one's understanding of the Bible.

Taking the Scriptures seriously involves affirming its status as Word of God and covenantal canon. It involves respecting the multifaceted character of its content, its sources, and its forms. It involves promoting its primary function of establishing a right covenantal relationship between the Creator and his creatures. And for us at this point, it calls for a disciplined study of the opening chapter of the Bible, the creation narrative in Genesis 1:1–2:3.

In making this study, we will have occasion to apply a number of the principles that we sought to establish in earlier chapters. Wherever it seems appropriate, we will employ the vehicle-packaging-content model for scriptural interpretation. We will recall the historical, religious, and cultural context in which the Bible was written to help us understand both the form and content of Genesis 1. Finally, we will read the Bible's first chapter in the context of the whole of Scripture.

---

1. For an example of this view, see Louis Berkhof's *Manual of Christian Doctrine* (Grand Rapids: Eerdmans, 1953), pp. 95-110.

One of the important hermeneutical principles that was reemphasized in the Protestant Reformation is the principle of interpreting Scripture by Scripture. This does not mean, of course, that the Bible should ever be interpreted in isolation from the rest of God's revelation in the Creation or in the words or deeds of his creatures. Rather, it means that because of the unity and integrity of Scripture, the whole provides insights into the understanding of its parts.

How might this principle of interpretation aid us in understanding the creation narrative of Genesis 1? If from an awareness of the structure and form of the whole Bible we have learned what role Genesis 1 plays in the canon, we will better understand what questions are being addressed. If we have gained from a familiarity with the whole of Scripture a sense of what kinds of questions about the material world are appropriate, we will avoid the tragic consequences of bringing inappropriate questions to Genesis 1. And if we have learned from a knowledge of biblical literature to recognize the characteristic forms in which the content of Genesis 1 is presented, we will be less likely to misunderstand its answers to appropriate questions.

It is, in my judgment, the failure to apply these principles to the reading of the creation narratives that has led to the blind alley of "scientific creationism," which is doubly misnamed, for it is neither consistently scientific nor biblically creationistic. But the contemporary revival of recent special creationism is a topic best left to a more complete discussion in a later chapter, so we will drop it for now and proceed with our study of Genesis 1.

Our goal in this chapter is to let God speak to us through Genesis 1, to allow Genesis 1 to serve as the vehicle of God's word to us, to see Genesis 1 as a vehicle filled with the artfully packaged content of God's message to us as it was to his people three millennia ago. We wish to "learn the wisdom that leads to salvation," and to "become fully equipped and ready for any good work" (2 Tim. 3:15, 17). We seek to understand Genesis 1 in the context of the whole of Scripture and in the context of all of God's acts in word and in deed, in history and in nature. In other words, we wish to take Genesis 1 seriously.

To accomplish this goal we will consider first the canonical function of the history portrayed in Genesis 1-11. We will then look into the genre of historical literature that this represents. And finally, we will explore the content of this vitally important opening chapter of Scripture.

GENESIS 1 AS HISTORICAL PROLOGUE

When I underscored in Chapter One the fact that the Bible has a very special status, I expressed its extraordinary character in two ways, characterizing it as "Word of God" and as "covenantal canon." Let us now develop the second term a bit more fully. As Meredith Kline cogently argues, it is essential for the proper interpretation of the Bible that it be recognized as a covenantal document—covenantal not only in *function,* but also in *form.* The form of scriptural material has remarkably close parallels in ancient Near Eastern international treaties used in the second millennium B.C.

> In this treaty form as it had developed in the history of diplomacy in the ancient Near East a formal canonical structure was, therefore, available, needing only to be taken up and inspired by the breath of God to become altogether what the church has confessed as canon. And that is what happened when Yahweh adopted the legal-literary form of the suzerainty covenants for the administration of his kingdom in Israel.[2]

The suzerainty treaties to which Kline is referring were binding written agreements between a suzerain (a great king with vast authority) and his vassals. Covenant documents discovered through archeological exploration ordinarily contain six characteristic elements: (1) a preamble that establishes the identity of the suzerain; (2) a historical prologue that reviews the relationship between the suzerain and his vassals, paying particular attention to the benevolent acts of the suzerain for which the vassal should be grateful; (3) the stipulation of the vassal's obligation for exclusive loyalty to the suzerain; (4) a policy for the safe keeping and periodic public reading of the covenant; (5) an appeal to witnesses (usually the natural or local deities); and (6) blessings and curses—statements describing the consequences of obedience and disobedience by the vassal.[3]

Our first step toward a better understanding of Genesis 1 must be to ascertain its role, or function, in the covenantal canon. Both its placement and its subject matter make it eminently clear

---

2. Kline, *The Structure of Biblical Authority* (Grand Rapids: Eerdmans, 1975), p. 37.

3. On the characteristics of suzerainty treaties, see Bernhard W. Anderson, *Understanding the Old Testament,* 3d ed. (Englewood Cliffs, N.J.: Prentice-Hall, 1975), p. 89; and G. Ernest Wright, *Biblical Archeology,* rev. ed. (Philadelphia: Westminster Press, 1962), pp. 100-101.

that it functions as a preamble identifying God as the Creator and as a segment of the historical prologue establishing the nature of the relationship between God and humanity. "If the Pentateuch is viewed as a unified corpus, with God's covenant with the exodus generation as its nucleus," says Kline, "the narratives of Genesis and the first part of Exodus assume the character of an historical prologue tracing the covenantal relationship to its historical roots in Yahweh's past dealings with the chosen people and their patriarchal ancestors."[4] Thus the book of Genesis forms part of the historical prologue to the covenant between Yahweh and Israel at Sinai. As such, its function must be respected. We should ask of Genesis neither more nor less than what it was designed to provide.

The book is further subdivided into two distinct parts: patriarchal history (chapters 12-50) and primeval history (chapters 1-11).[5] Patriarchal history, drawn from the past as "remembered" in oral tradition, traces the creation and election of the nation of Israel back to the promise (covenant) made by Yahweh to Abraham. The covenant at Sinai is thus placed in the context of a long-standing relationship between God and the seed of Abraham. Primeval history, drawn from the past as constructed from literary and religious tradition, places the divine call to Abraham in the context of the timeless relationship of God to all of humanity and to the whole cosmos—celestial, terrestrial, even subterranean. Genesis 1 forms the magnificent preamble and opening narrative in this primeval history that introduces the God of the covenant, the God of Abraham and Moses.

---

4. Kline, *The Structure of Biblical Authority*, p. 53.

5. Any number of commentaries on Genesis discuss its structure and significance. I will note a few that I found most informative and helpful. For a thorough scholarly treatment, see Gerhard von Rad's *Genesis: A Commentary*, trans. John H. Marks, rev. ed. (Philadelphia: Westminster Press, 1973). For an excellent and highly readable commentary from a Jewish perspective, see Nahum M. Sarna's *Understanding Genesis* (New York: Schocken Books, 1970). For commentaries on just the primeval history section, see Alan Richardson's *Genesis 1-11*, Torch Bible Commentaries (London: SCM Press, 1953), or vol. 1 of John C. L. Gibson's *Genesis*, The Daily Study Bible Series (Philadelphia: Westminster Press, 1981). For a discussion of just the first three chapters of Genesis, see Claus Westermann's *Creation*, trans. John J. Scullion (Philadelphia: Fortress Press, 1974). Two recent publications that contribute particularly helpful perspectives are Henri Blocher's *In the Beginning: The Opening Chapters of Genesis*, trans. David G. Preston (Downers Grove, Ill.: InterVarsity Press, 1984), and Conrad Hyers's *The Meaning of Creation: Genesis and Modern Science* (Atlanta: John Knox Press, 1984).

GENESIS 1 AS PRIMEVAL HISTORY

The call and promises made by Yahweh to Abraham, as recorded in Genesis 12:1-3, ought to arouse a profound curiosity in the reader, whether ancient or modern. Surely one would be led to inquire who this Yahweh is who gives both commands and promises to Abraham. What are his credentials? What is his authority? What are his powers? And what is his status? Where does Yahweh stand among the numerous deities of nature and nations? In Abraham's world such questions would be most appropriate.

Furthermore, questions concerning the identity and status of humanity would also have come to the mind of the ancient reader (or listener). What is the nature of mankind? What is humanity's status and relationship to Yahweh? What is a person's relationship to other people and institutions? Why is there evil and suffering in the human experience? Why are there divisions among persons and nations? What is man's responsibility to Yahweh? To other people? To nations?

Finally, the readers' legitimate curiosity would question the identity and character of their habitat, the natural world in which they lived and that they so intimately experienced. (We twentieth-century Western people are usually so insulated from the threat of domination or destruction by the material world that we easily overlook the fact that this was a persistent concern of our ancient predecessors.) In the context of their intimate awareness of the power of natural forces, the ancient Hebrews would likely have wondered about the character of the world in which they lived, about its powers, about its status relative to God and to humanity, about their place in it and their responsibility toward it.

These questions and many others like it are the subject matter of the primeval history in Genesis 1-11. Primeval history sets the stage for God's entering into a covenant relationship with Abraham in ancient Mesopotamia. In a brief and compact prologue to patriarchal history and the covenant at Sinai, the primeval history in Genesis 1-11 provides God's answers to many of humanity's most profound questions. The questions addressed by primeval history are universal questions—questions that are asked by people of all nations, all cultures, in all times. They are our questions just as much as they were ancient Hebrew questions. We stand as much in need of the answers as the Israelites did in the desert at the foot of Mt. Sinai.

Within the complete covenantal canon, the book of Genesis functions as part of the historical prologue to the Mosaic covenant stipulations. Within the book of Genesis, the first eleven chapters form a body of primeval history that sets the stage for Abraham's experience of being called by God and promised both land and descendants. Within the primeval history section, Genesis 1, following the suzerainty treaty pattern, functions primarily as the preamble that establishes the identity of the suzerain and the relative status of suzerain and vassal. As such, Genesis 1 shows not the slightest interest in natural science, ancient or modern. Thus, neither ancient Babylonian mathematical astronomy nor twentieth-century astrophysics will shed much light on the message of Genesis 1.[6]

A proper interpretation or understanding of Genesis 1 requires not only the correct identification of the questions it addresses but also a familiarity with the form or structure of its answers. The preamble addresses essential questions of identity and relationship, central among which is the question of the identity of God and the issue of how he is related to humanity and our natural environment. Primeval history is the vehicle that conveys the answers to these profound questions. But how are the answers packaged? In what form does primeval history present its answers? Failure to recognize or to respect the manner in which Scripture's message-content is packaged has provided (and, unfortunately, will probably continue to provide) numerous occasions for misunderstanding. For a large number of Christians, Genesis 1 appears to provide a particularly difficult challenge.

The difficulty lies, I believe, in the fact that the form of primeval history's answers to mankind's fundamental questions of identity is markedly different from the form we are most familiar with and that we consequently expect. As persons born and educated in the context of twentieth-century Western culture, we have been trained to think and communicate in patterns drawn largely from Greek civilization. One of the greatest contributions of Greek culture was the development of

---

6. For an interesting though essentially misconceived attempt to interpret Genesis 1 in the light of modern astrophysics, see Robert C. Newman and Herman J. Eckelmann, Jr.'s *Genesis One and the Origin of the Earth* (Grand Rapids: Baker, 1977). In my judgment, the work's concordistic approach addresses entirely inappropriate questions to Genesis 1 and diverts attention from the fundamental questions of primeval history.

abstract discourse. It is largely because of the Greek development of abstract thought that we can, for example, sensibly think and speak of the quality of beauty without being limited to the consideration of specific or concrete beautiful objects.

Primeval history, however, was not written in the Greek, Western manner. The vast majority of the Old Testament was written prior to the development of Greek philosophical culture or its influx into Palestine. Primeval history is Hebrew literature written in the literary tradition of ancient Near Eastern cultures. Hebrew literature displays little abstraction but instead relies almost exclusively on concrete illustration to make a point. The most common literary genre in the Old Testament is the narrative—the story in simple, straightforward prose form. The qualities of God, humanity, and nature are not discussed abstractly in philosophical exposition but rather are illustrated by stories of events and actions carried out by God, humanity, and nature. As characters in a story, God, humanity, and nature perform specific actions that illustrate their identity, their character, and their relationships. The stories serve as "packaging" that contains the message-content conveyed by the vehicle of primeval history.

But if the stories of primeval history are the packaging rather than the content itself, ought they still to be taken seriously? Of course! We must find in them the answers to the questions addressed by primeval history. We must discover what qualities of God, humanity, and nature they are illustrating by means of concrete story elements. The stories of primeval history are to be taken very seriously because they convey to us the answers to some of the most profound questions ever asked by humanity.

Yet I know that there is a further question that Christians desire to have answered: Are the stories true? There is no question that they are true in the sense that they illustrate and convey truths about the identity of God, humanity, nature, and their relationships to one another. It is the purpose of primeval history to answer precisely these questions, and by faith we believe that the scriptural answers are true. However, most twentieth-century Westerners are more specifically interested in whether the events actually happened just as they are reported in the narratives. But that is a Western question, not an ancient Eastern or Hebrew question. It shifts the emphasis away from the heart of the matter and directs attention to peripheral matters, to matters beyond the scope of the narrative. It sidetracks our train of thought away from the main line and sends us down a dead-end spur.

The truth of a concrete story in ancient Hebrew literature does not necessarily lie in its specific details but rather in the eternal verities it illustrates. When we modern Westerners read a story, we expect it to be written as an answer to the question "What happened?" But the stories of primeval history are much more like parables than like journalistic reports of events. They illustrate the identity and character and status of God, humanity, and nature. They were never intended to answer questions about precisely what happened; rather, they were designed to answer questions about the character of the chief participants in the human experience and the nature of their relationship. In typically Eastern fashion, primeval history answers these questions with illustrative stories that share many features with the parables we find elsewhere in Scripture.

Primeval history and parable can both serve as vehicles of truth—important truth. In both cases, the concrete details of the story constitute the packaging in which that truth is conveyed. In both cases the content of truth is of infinitely greater value than the vehicle or packaging in which it is carried. In either case, if we attempt to consume both the content and the packaging, we may encounter significant difficulty in chewing, swallowing, and digesting the combination. Those who want to feed on the truths of Scripture must take care to differentiate between food and packaging.

Unlike parables, primeval history does refer to a historical past with a character essentially the same as that illustrated by the narrative. Though actual history and the primeval narratives may differ vastly in detail, they belong to the same genus. Primeval history is not simply early history or prehistory; it is a collection of narratives that provides the conceptual framework necessary to understand all of history. Primeval history sets the stage on which actual human history is played out; it provides the framework in which history is to be experienced. Its stories apply not merely to specific events or individuals but to all of history and all of humanity. Primeval history is as much our experience as it is anyone's. Primeval history tells us as much about our God, our selves, and our world as it tells about the God, the person, and the world of Abraham, or Moses, or David long ago. Though it is not to be taken literally, it is to be taken seriously.

Genesis 1 is perhaps the grandest of all primeval histories.

In a simple yet elegantly structured creation narrative, Genesis 1 provides us with the key to understanding all of history: God is the Creator, and we, along with the whole cosmos, are his creatures. Genesis 1 is a story. It is commonly called the "story of creation"—although in light of its relationship to the whole Bible, in light of its role as primeval history, and in light of its function as preamble to the covenantal canon, I would judge it far more accurate to call it the "Story of the Creator." Its purpose is not to chronicle the divine creative process but to identify the God of the covenant as the Creator of both man and his habitat.

Thus, the proper question to bring to Genesis 1 is "Who is God and how are man and the world related to him?" The answer is given in the form of a story that illustrates the identity of God and his relationship to humanity and the cosmos. But the story so vividly portrays its action that we are irresistably led to wonder about the chronology of the narrative. In the story, God the Creator is clearly portrayed as performing his creative works within a six-day period and resting on the seventh. What must we make of that chronology? What does the seven-day structure signify?

The first point we should note is that, compared with the principal message of Genesis 1, matters of chronology and timetable are decidedly secondary in importance. We may have an intellectual curiosity about these matters, and we may praise God the Creator that we live in a day when that intellectual curiosity can be at least partially satisfied, but we must recognize that questions of chronology are not pivotal. And we must recognize that questions of chronology beyond the limits of the human experience, whether past or future, whether "in the beginning" or at "the end of time," are *not* the subject of the biblical message. The beginning lies shrouded in mist beyond human memory, and the end will come "as a thief in the night."

The seven-day chronology that we find in Genesis 1 has no connection with the actual chronology of the Creator's continuous dynamic action in the cosmos. The creation-week motif is a literary device, a framework in which a number of very important messages are held.[7] The chronology of the narrative is not the chronology of creation but rather the packaging in which the message is wrapped. The particular acts depicted in the Story of

---

7. See N. H. Ridderbos, *Is There a Conflict between Genesis 1 and Natural Science?* trans. John Vriend (Grand Rapids: Eerdmans, 1957).

the Creator are not the events of creative action reported with photographic realism but rather imaginative illustrations of the way in which God and the Creation are related.

GENESIS 1 AS A REVELATION OF GOD

If Genesis 1, functioning as a preamble and prologue in the covenantal canon, employs the vehicle of primeval history and the packaging of a seven-day cycle of work and rest, what is the content that is being conveyed? What message does it deliver? What does it reveal to us, as it did to the ancient Hebrews, about our God? After all of this preliminary discussion about vehicle and packaging, is there anything left to identify as content?

With confidence and enthusiasm we can say Yes, there is indeed a substantive message conveyed by Genesis 1. It is one of the two most significant revelations ever made to humanity. It proclaims in no uncertain terms that God is the Creator and that the whole cosmos is his Creation. (The second revelation essential to humanity is that God is the Redeemer and that the redeemed are his new creation.) Let no one say that identifying Genesis 1 as primeval history and the seven-day creation week motif as a literary framework reduces the importance or the impact of its contents. On the contrary, I firmly believe that an honest recognition of and respect for these canonical and literary qualities, arising appropriately in the historical and cultural milieu of the ancient Hebrews, allows us to perceive the true message of Genesis 1 with greater force and clarity than ever before was possible.

The fundamental question addressed by Genesis 1 is "Who is the God who called Abraham, and how is he related to humanity and the natural world?" The answer, so beautifully and effectively illustrated in narrative form, is that God is the Creator of the heavens and the earth and all of their inhabitants, both celestial and terrestrial. And beyond that fundamental revelation, Genesis 1 provides the basis for the full biblical revelation concerning what kind of a Creator our God is and what it means to be a creature who is covenantally related to him.

The Creator who is introduced to us in Genesis 1 is transcendent—he is not equated with the cosmos or any part of it but is presented as a being whose essence transcends the limitations of the corporeal. This Creator is not merely a personification of some natural object; he is revealed to be a spiritual person who called the whole corporeal cosmos into existence and who continues to sustain its existence moment by moment. The Creator that Genesis 1 presents to us is the transcendent Originator and Preserver of the whole cosmos. He is the One who has the power and authority to say "Let there be," and it is called into existence. But remember, this is a statement introducing and describing the Creator, not specifying the manner or timetable for such creative activity. It employs anthropomorphic language to illustrate the character of the Creator; it does not employ ancient scientific language to report the mechanism of the creation process.

The Creator introduced to us in Genesis 1 is not only transcendent but also sovereign: he is the supreme ruler or governor of the cosmos. He is the one who directs his Creation to have an ordered structure and a patterned behavior. The giving of ordered structure is illustrated by the familiar divisions of light from darkness, the waters above from the waters below, and dry land from the sea—common ancient Near Eastern concepts of cosmic structure. While structure may appear to be a rather static manifestation of divine governance, patterned behavior or development constitutes a dynamic aspect of the Creator's governing activity. Already in Genesis 1, this dimension of God's immanent and dynamic interaction with his Creation is introduced. The earth produces vegetation, plants bear seed, sea creatures and birds multiply, the earth produces animals of all kinds—all familiar activities and processes are said in Genesis 1 to be responses to the Creator's direction. The Creator introduced to us in Genesis 1 is the sovereign Governor of all things—every part of the cosmos has an ordered structure and a patterned behavior as a response to the Creator's governing power.

According to Genesis 1, the Creator is not only the transcendent Originator and Preserver of the cosmos and the immanent and sovereign Governor of all structure and behavior of the corporeal world but also the benevolent Provider. The Story of the Creator portrays God as one who recognizes the needs of his creatures and makes provision for them. Seed-bearing plants and trees supply food for mankind. Foliage is given to feed animals and birds. What a splendid way to be introduced to the God of the covenant. He is the One whose goodness can be experienced each time our hunger is satisfied by food.

Without doubt, the central figure in the Genesis 1 story is the Creator; after all, the chief question addressed by this narrative is the question concerning the identity of the covenant God. But there are other figures in the story as well—the material world and mankind. What can be learned from Genesis 1 about the cosmos?

Against the background of ancient Near Eastern naturalistic polytheism, Genesis 1 says some startling things about our natural environment. According to ancient pagan religions, the natural world was a manifestation of numerous deities—gods of sky, sun, moon, stars, storms, mountains, rivers, and seas. The gods of polytheism were personifications of objects and forces in nature. Or, to look at it from the other side, the cosmos had the status of deity. To this concept Genesis 1 says "Nonsense!" The cosmos is neither God nor an assembly of gods. The cosmos does not have the status of deity; it has the status of Creation. Everything—light, sky, dry land, seas, vegetation, celestial luminaries, fish, birds, animals, man, and woman—every being in the heavens and on the earth stands under the one and only God as his Creation. Nothing in this world has the status of deity. That status belongs solely to the transcendent, sovereign Creator. The relationship of God to the world is the relationship of Creator to his Creation.

The designation of the status of the world as Creation in Genesis 1 has additional implications. As Creation, the cosmos has an existence that is radically contingent—entirely dependent on both the originating and the sustaining action of the Creator. As Creation, the corporeal world exhibits a patterned behavior only as a response to the Creator's governing power; material behavior is not self-willed but God-governed. Genesis 1 not only strips the world of divine status but divests natural forces of any personal character.

What about value and purpose? I think that already in Genesis 1 we find the seeds of Scripture's teaching that the material world has no value in itself but only in its relationship to the Creator. And although we find purposeful relationships and events throughout the narrative, it is clear that their purpose derives not from any thing or phenomenon itself but from the grand purposes of the Creator. Value and purpose are not inherent in the cosmos; they derive solely from its relationship to the Creator.

We should also note that humanity is assigned a special status in the Genesis 1 story. We are a part of Creation, but we alone are said to bear God's image; that is to say, we have been created with both the capability and responsibility to serve the Creator in a unique way—the way specified in the covenantal agreement. That opens up an area of consideration far beyond the scope of this discussion, however, so let us restrict our focus once again to the celestial luminaries.

The Fourth Day of Creation

> *Then God said, "Let there be lights in the expanse of the heavens to separate the day from the night, and let them be for signs, and for seasons, and for days and years; and let them be for lights in the expanse of the heavens to give light on the earth"; and it was so. And God made the two great lights, the greater light to govern the day, and the lesser light to govern the night; He made the stars also. And God placed them in the expanse of the heavens to give light on the earth, and to govern the day and the night, and to separate the light from the darkness; and God saw that it was good. And there was evening and there was morning, a fourth day.*
>
> —Genesis 1:14-19, NASB

This episode in the creation narrative focuses principally on the practical functions served by the "lights in the expanse of the heavens." They establish the cycle of day and night, they provide light for the earth, and they serve various calendric functions. Mention of these matters ought to come as no surprise. The regular motions of sun, moon, and stars have provided a means for timekeeping throughout recorded human history. As we noted in the historical overview in Chapter Two, the Mesopotamians developed an extremely sophisticated mathematical astronomy during the Old Testament period. The repeated references in Genesis 1 to the practical functions served by the heavenly lights—the provision of light and the marking of calendric intervals—suggests that these phenomena are to be enjoyed and appreciated as gifts from the hand of the Provider.

But beneath the surface of this recounting of the practical services performed by the celestial lights there lies a penetrating polemic against the naturalistic polytheism of Israel's neighbors. The sun and moon, for example, are not gods to be worshiped; they are simply parts of the Creation—no more, no less. Lest even a reference to their names suggest the pagan deities associated with them, Genesis 1 refers to them simply as the "greater light" and the "lesser light." Furthermore, the stars (and planets) are not astral deities to be feared; they are parts of God's good Creation. And lest they be given too prominent a position, Genesis 1 treats them as little more than an afterthought, adding "He made the stars also."

In sum, Genesis 1 teaches us a number of very important things about sun, moon, and stars. We learn that they have the status of Creation. We learn that stars, like all things in the Creation, are totally dependent on God for their existence. We learn that their material behavior is governed by the power of the Creator. We learn that their value is not intrinsic but resides in their relationship to the Creator. We learn that the purposes they serve are not their own but that they exist to serve the purposes of the Creator within the whole Creation and for the good of all of his creatures. In other words, we learn the answers to important questions concerning the status, origin, governance, value, and purpose of stars—all features of the relationship between the Creation and the Creator who is being introduced by Genesis 1.

But there are many other questions about the heavenly bodies and the Creator's acts that are often addressed to Genesis 1. Let us briefly consider a few of these.

GENESIS 1 AND SCIENTIFIC QUESTIONS

On my bookshelves are dozens of books dealing with the topics of creation, natural science, and the relationship between them. And when I visit my local religious bookstores, I find whole shelves filled with books in the category of "Bible and science." A few of them are worth reading; many more are not. The majority of books on this topic in today's market have been written under the assumption that Genesis 1 must be treated as reportorial prose, as a chronicle of distinct, instantaneous creative acts performed in exactly the sequence stated in Scripture and within a period of six ordinary twenty-four-hour days, approximately ten thousand years ago. That assumption raises a considerable number of difficult questions of an essentially scientific nature, however—questions pertaining to the material properties, physical behavior, and temporal development of the universe.

Light before Lights?

Within the creation-week scenario, we find the creation of light on day one, but the sources of light—the sun, moon, and stars—do not appear until the fourth day. How is it possible to have light before lights? My imaginative powers are put to shame by the multitude of clever conjectures and adroit arguments that have been fabricated to solve this puzzle. But the puzzle should not be solved. It should instead be summarily dismissed. It needs no solution because the question itself is unwarranted. The question presupposes that the order of events in the Story of the Creator has some physical basis. It does not.[8] It may have a cultural basis, but surely not a physical or material basis. Bringing a question of chronological order to Genesis 1 is like bringing a question of meteorology to Psalm 139. Both efforts would simply generate the sort of meaningless speculations that inevitably result from attempts to press inappropriate questions on Scripture. Bring nonsense questions and you will get nonsense answers.

Solar Days before Sol?

In the same category as the question of light before lights is the question of the length of creation days before the appearance of the sun (Sol). The repeated phrase, "and there was evening and there was morning, a . . . day," quite naturally suggests an ordinary solar day of twenty-four hours. After the appearance of the sun, the twenty-four-hour cycle would be established in the usual manner—the earth's daily rotation relative to a line from the earth to the sun. But before the sun was present as a light source to "govern the day," what would have been so special about a twenty-four-hour period? In the context of this ambiguity, some Christians have been led to suppose that the first three days of creation had a different length than the fourth and subsequent days.

Genesis 1 gives us no clues to the solution to this riddle—and for good reason. The riddle presents to us a nonsensical question, a "nonquestion." The days of Genesis 1 are a literary device: they are story elements, not temporal specifications. Their relative lengths are a matter of no significance whatsoever. To waste time worrying about such matters is poor stewardship of our mental energies. Nonsense questions generate only nonsense answers.

---

8. Ridderbos, *Is There a Conflict,* p. 68.

How Many Years in a Day?

Because the physical evidence, both geological and astronomical, so strongly indicates that cosmic history extends over a period greater than ten billion years, many sincere Christians have attempted to interpret the days of Genesis 1 as long periods of creative activity rather than as literal twenty-four-hour solar days. The advantage of this interpretation is that it acknowledges the validity of the empirical evidence while removing the appearance of a contradiction between the biblically derived and scientifically derived cosmic chronologies. Some hold that such an approach produces a concord, a harmony, between the two means of establishing the timetable of God's creative activity.

To evaluate this concordistic approach, let us begin by noting that it both asks a question and provides an answer. The question is how long the "days" of creation lasted. The answer required to achieve a harmony between the empirically derived chronology and a chronological interpretation of Genesis 1 is that each "day" must be some two to three billion years long. How many years in a day? Billions!

But again, I would protest that the question is itself irrelevant or meaningless. The days of the Genesis 1 story are clearly ordinary solar days. There is nothing in the story itself to indicate that they should be thought of as any other time period. All of the discussion that one finds in the literature about the various uses of the Hebrew word *yom* ("day") for time intervals other than twenty-four hours may be interesting, but the discussion is in this context entirely irrelevant. The days of Genesis 1 have nothing to do with the cosmic timetable; they are simply literary devices in the story, not actual temporal intervals directly corresponding to events in cosmic history. Extending their length to several billion years in order to accommodate the scientifically derived measure of the span of cosmic history only creates a new set of unresolvable problems.

The concordistic interpretation is another example of the meaningless explanations that derive from addressing inappropriate questions to Scripture. It is the sterile offspring of the unwarranted assumption that Genesis 1 is meant to be a journalistic recounting of God's original creative activity, a log of God's specific acts in constructing the cosmos and the creatures that inhabit it.

Any chronological interpretation of Genesis 1, whether it makes the assumption that its temporal units are solar days, geologic ages, or multibillion-year eons, is doomed to failure. Any interpretation presuming that the chief question addressed by this narrative is "What happened?" has failed to employ nearly all of the interpretive principles that we have sought to establish. An interpretation that fails to identify the canonical function of Genesis 1 as preamble and prologue to the covenant will fail to promote the primary function of Scripture—namely, to bring the reader (or hearer) into a proper covenantal relationship to God. An interpretation that fails to recognize that the genre of Genesis 1 is primeval history will fail to respect the diversity of Scripture's literary forms, each requiring its own unique interpretive methodology. And, by failing to read Genesis 1 in the context of the whole of Scripture, a reader will fail to identify correctly the questions that are being addressed there.

In place of the fundamental question of the covenantal prologue—"Who is the God of Abraham and Moses?"—both the concordistic and literalistic interpretations substitute the semiscientific question "By what mechanism and in what time frame did God construct the world?" Both of these chronological interpretations treat the seven-day structure of the Genesis 1 narrative as if it were a temporal specification rather than a literary framework, and in so doing, they fail to distinguish between the content and packaging of Scripture, between the story elements and the message being conveyed by the story.

Insisting on a chronological interpretation of Genesis 1 not only directs our attention to the wrong question but also leads us to the wrong source for answers. Having raised the question of how long the days of Genesis 1 are, concordists then go to the natural sciences for answers. But this means that they are asking of the sciences, which by their nature deal only with the material world as a thing in itself, a question concerning divine activity, a question on which they are unable to speak. The natural sciences can provide an answer to a question about what empirical investigation reveals about cosmic chronology, but neither geology nor astronomy can answer questions about the length of the days of Genesis 1. To treat the empirically discovered cosmic chronology as if it were the answer to the question of Genesis chronology is to allow the sciences to dictate scriptural interpretation, which is, in general, bad hermeneutics. But that is what often happens when inappropriate questions are directed to the Bible. If answers are demanded, the answers, along with the methodology of generating them, will be equally inappropriate.

Lest the reader get the impression that I judge questions of cosmic chronology to be either unimportant or uninteresting, let me say explicitly that the truth lies in precisely the opposite direction. Our God-given curiosity leads us to inquire into such matters. If the cosmos is God's Creation, as the Scripture so clearly teaches, then its history is intrinsically significant. And if the behavior of the cosmos is divinely governed, as the Bible so vividly reveals, then the temporal development of the universe and the creatures that inhabit it is worthy of our most diligent and honest investigation. To state it even more strongly, because God is the Creator, and because the cosmos is his Creation, we (his image-bearing creatures) are obligated to study the Creation diligently as a means of knowing the Creator and thereby preparing ourselves better to serve and praise him as our covenant God.

If we accept this responsibility and choose to investigate the properties, behavior, and chronology of the Creation, we will have to turn to the Creation itself as a source of information, and to the natural sciences—employing both empirical observation and theoretical analysis—for our methodology. To find the answers to questions about the physical properties, the material behavior, and the chronological development of stars, we must go not to ancient Palestine, but to Palomar. We must scan the skies with the aid of the modern observational instruments of an astronomical observatory. The spectacles of Scripture enable us to perceive the status, origin, governance, value, and purpose of stars—all matters of their relationship to the Creator. The telescopes of modern astronomy give us the opportunity to look into the properties, behavior, and history of stars—all matters of their qualities as things in themselves. Without removing our spectacles, let's now spend some time peering through the telescope.

PART II:

The Scientific View

The 200-inch-diameter telescope of the Hale Observatories, located on Palomar Mountain in southern California.

CHAPTER SIX

Taking the Cosmos Seriously

In the preface I remarked that this work is directed to readers who want to take both the Bible and the material world—the cosmos—seriously. To this point our principal focus has been on what it means to take the Bible seriously. Now we can turn to the question of what it means to take the cosmos seriously.

There are, of course, any number of ways in which to approach the issue of what it means to take the cosmos seriously. Since we are specifically concerned to learn something about the properties, behavior, and chronology of stars through astronomical investigation, we will direct our question to the natural scientist—in this case an astronomer. Having been engaged in the practice and teaching of physics and astronomy for the past two decades, I will formulate an answer based on my own training and experience. I trust that it will be representative of my profession and that it will provide a useful summary of the character of natural science as it is presently defined and practiced.[1]

1. For the purposes of our discussion of the nature of scientific investigation, I have chosen to take what might be called a "philosophically naive" approach. Rather than present an analysis of the scientific enterprise from a formal philosophical perspective, I have chosen to present a view of science as seen from the inside. Thus, I have not drawn heavily on the analysis or vocabulary of philosophers of science but have instead drawn from the attitudes and viewpoints of teachers and colleagues who are engaged in the practice of science itself. For those who wish to examine the philosophical perspective, I recommend Harold I. Brown's *Perception, Theory and Commitment* (Chicago: Uni-

In my judgment, the answer to the question of what it means to take the cosmos seriously must incorporate several elements: it must provide a statement of the *assumptions* that a natural scientist makes concerning the material world, it must define the *domain* of scientific investigation, it must specify the *methods* to be employed in the study of the material world, and it must discuss the nature of the *results* of the scientific enterprise. To these four matters we now proceed, not aiming at anything approaching an exhaustive discussion of these weighty topics, but in hopes of assembling a sufficient basis for understanding the nature of both the quest and the discoveries of modern astronomy and astrophysics.

THE ASSUMPTIONS OF NATURAL SCIENCE CONCERNING THE MATERIAL WORLD

On the Character of Its Existence

Most practicing scientists make essentially commonsense assumptions about the character of matter's existence. We assume that the material world has an objectively real existence—that is, it exists independent of our perception of it. We assume that the cosmos is more than a mere product of our imagination. Stones and stars, galaxies and geese, really exist, whether or not we happen to be observing them or thinking about them.

We also assume a continuity of existence in the cosmos—that is, that each state of the universe, or of individual objects in it, is intimately related to states immediately preceding and following it. Though the cosmos is constantly changing, what we have after a change is not a new cosmos but the old one in a new state or form. Modern science suggests that on the microscopic level this assumption of continuity encounters some interesting difficulties, but on the macroscopic scale the concept of continuity of existence to which common sense and ordinary experience would lead us appears to be quite sound. I rest assured, for example, that the desk at which I am seated has existed continuously during the past several decades and that the numerous nicks and scratches on it reveal the character of many past events in which it, as a continuously existing object, has participated. The presupposition of continuity in existence provides the

---

versity of Chicago Press, 1977). Numerous references to associated literature can be found in that work.

foundation for the study of history, whether of desks or nations or the universe.

What about the cause for the existence of the cosmos or any object in it? Does the universe have a radically contingent, externally caused existence, or does it have an independent, self-caused existence? What assumption does natural science make concerning the cause or ground of existence? Quite simply put: none. Questions concerning the cause for the existence of the cosmos are important and fascinating questions, but they cannot appropriately be addressed to natural science. They are metaphysical questions that must be addressed to philosophy or theology; the natural sciences are simply not equipped to handle them.

Similarly, no presupposition can be made by the natural sciences concerning the status of the material world relative to any nonmaterial entities. Natural science, by the definition of the discipline and by the limitations of its observational tools, knows of no nonmaterial entities and thus is simply unable to speak on matters of their status or relationship. It knows of matter and the space-time continuum in which it functions, but it can say nothing about status—nothing positive, nothing negative, just nothing.

On the Character of Its Behavior

Perhaps the most general statement that we could make about the modern scientific study of the cosmos is that the natural sciences assume that the behavior of matter is far more like the workings of an impersonal machine than like the actions of a willful personal agent.[2] It is this assumption or attitude that

---

2. R. G. Collingwood has argued in *The Idea of Nature* (Oxford: Oxford University Press, 1945) that the mechanical view of nature, based on an analogy with machines, has been replaced by a view based on an analogy with history. It seems to me, however, that in recent decades the idea of the cosmos as an impersonal machine has been revived, perhaps as a result of the progress in molecular biology, which suggests to some that even living creatures may be viewed as no more than "marvelous molecular machines."

It is true, though, that contemporary natural science views the cosmic machine differently than did the scientific community of a century or two ago. The classical physics of that era had a very deterministic concept of cause and effect. It was maintained that if each cause can have only one prescribed effect, then the future of the entire universe would have to be fully determined by its state at any instant. Contemporary science recognizes that the cause-effect relationships exhibited by material behavior include the possibility of there being

provides the fundamental contrast between the modern scientific age and the era of primitive animism. In our day, the scientific enterprise views the cosmos as an insentient machine that behaves according to rigorously set patterns. In the era of animism (not entirely past) the cosmos was viewed more like a sentient person that freely and consciously chose to act as it did.

So, natural science assumes that the cosmos and all of its parts behave according to certain patterns, something like a machine the motions of which are governed by its design and fabrication, or like a computer governed by its program. But what sort of patterns are they? As I reflect on my own training in physics and astronomy, certain important underlying (often unstated) assumptions come to mind. As a general rule, we assume the orderly, patterned behavior of matter to be spatially and temporally invariant, coherent, and causally related to the properties of matter itself.

By "spatial invariance" I mean that matter behaves according to the same rules everywhere in space. We believe, for example, that atoms behave in the same way on the planet Venus as they do here on earth. Provided with identical environments, matter will behave according to identical patterns, irrespective of location. We assume (and there is abundant evidence to support this presupposition) that whatever agent is governing material behavior acts in the same manner everywhere.

By "temporal invariance" I mean that the rules for material behavior do not change with time. We assume, for example, that atoms behave according to the same patterns today as they did yesterday and a thousand years ago and for as long as atoms have existed. We assume (and there is abundant evidence to support this assumption) that whatever agent is governing that behavior, whether internal or external, governs it in the same manner at all times.

---

any number of effects following from a given cause (although some effects may be less likely to occur than others, and indeed, it may be possible to predict the relative frequency of each possible effect with some considerable precision). Thus, even in a nondeterministic world the future states of which are contingent upon the particular sequence of effects permitted by the prevailing causes, there is still patterned material behavior. Though the material world is not deterministically predestined to follow a fixed sequence of states, it does exhibit behavior that is constrained to occur within the limits of certain requirements. According to contemporary natural science, the cosmic machine is more complex and less predictable than the machine conceived classically, but it is just as impersonal and insentient.

Furthermore, we assume that matter behaves according to patterns that are noncontradictory and coherent. We expect to find that the patterns, or principles, or "natural laws" that apply to one class of phenomena are consistent with the laws applicable to other categories of phenomena. For example, we assume that the principle of energy conservation (which we'll discuss more fully in the next chapter) applies to all material systems—to elementary particles, to atoms and molecules, to cells of living tissue, to organisms, to planets and stars, and even to galaxies. Such a principle is not arbitrarily obeyed or disobeyed by different assemblies of matter; rather, there is a coherence of behavior exhibited by all systems. We assume (and there is abundant evidence to support this assumption) that whatever agent governs material behavior does so in a coherent and noncontradictory manner.

Since the work of Isaac Newton in the seventeenth century, the natural sciences have been characterized by the successful search for causal relationships between the properties and the behavior of matter and material systems. For example, early in the seventeenth century, Johannes Kepler discovered the correct description for the orbital motion (an example of patterned behavior) of planets relative to the sun, but his description offered no answers to questions of cause. Kepler discovered that planets move in a certain way, but he could not explain why they should exhibit that particular behavior. It was later in that century that Newton formulated a system of thought about the motion of material bodies that provided answers to the questions of cause. At the heart of Newtonian dynamics is the idea that acceleration (any change in a body's speed or direction of motion) is caused by a force. Add to this cause-effect relationship for motion Newton's postulate concerning the gravitational force exerted by one body on another—by the sun on a planet, for instance—and the cause for the planetary motion described by Kepler could be identified: planets move as they do in response to the gravitational influence of the sun. Even more startling, the cause for the particular motions exhibited by the planets was identified as essentially the same as that which causes apples to fall earthward. The patterns and the causes of behavior in the celestial realm were discovered to be the same as those in the terrestrial realm.

On the strength of the repeated discovery of such universally applicable cause-effect relationships, natural science now works on the presupposition that all material behavior is characterized by some form of immanent causality. It is assumed that every effect requires a cause and that every cause will have a prescribed effect. It must be noted, however, that we are talking here only about immanent or proximate causes, not about transcendent or ultimate causes. The cause-effect relationships discovered or assumed by the natural sciences are perhaps better identified as statements of correlation than as explanations of cause. Newtonian mechanics, for example, provides a set of statements and a computational program suggesting, in essence, that if a system of material bodies has certain properties, then its behavior will proceed according to a certain corresponding pattern. There is, in other words, a strict correlation between a system's properties and its behavior. This can be legitimately called a "cause-effect relationship" only if one understands that the word *cause* is being used in the sense of a proximate or immanent condition that precedes the "effect" and not in the sense of an ultimate or transcendent power (or causer) that determines that such a condition will necessarily produce the observed effect. The science of physics allows us to say that the falling of apples to the ground is caused by the force of gravity. But why do massive bodies exert gravitational forces on one another? What is the ultimate source of that mode of interaction? These questions of ultimate or transcendent cause are appropriately addressed not to the natural sciences but to metaphysics or theology (a matter to which we will return later).

Within the limits of proximate and immanent causes, the natural sciences generally assume a continuity of cause-effect relationships. We assume that at any instant the state of the cosmos is causally connected with those states immediately preceding and following. Or, to put it more simply, today is a product of yesterday. Similarly, the state of affairs in the cosmos tomorrow will follow from the events of today.[3]

This continuity, something that we experience daily, is assumed to apply not only to those phenomena within memory or within the limited range of personal observation but to all phenomena in the material world. The patterned behavior of matter is assumed to form a historical continuity of causally related events. We assume that the present is but one moment in

3. Again, this is not to say that the future is fully determined by the present or that the present has been fully determined by the past. Because of the multiplicity of effects permitted by some causes, the continuity of cause-effect relationships does not imply that cosmic history is deterministically prescribed.

a continuum of moments that are causally connected. We assume (and there is abundant evidence to support this assumption) that whatever power governs the cosmos does so in a manner not only spatially and temporally invariant, not only consistent and coherent, but also causally continuous. Cosmic governance, we believe, is neither incoherent nor inconsistent nor discontinuous nor capricious nor devious.

But even if the behavior of the corporeal cosmos follows rigorous patterns that are invariant, coherent, and causally connected, the practice of natural science requires more; it requires that this behavior be perceptible to the human senses and intelligible to the human mind. On the one hand, this statement reflects an additional assumption that natural science makes concerning the qualities of material behavior. On the other hand, it is an admission that the natural sciences have access only to those dimensions of material behavior for which this assumption is satisfied. Imperceptible or unintelligible behavior will by its very nature be excluded from consideration; there is no alternative. If there are such dimensions to material behavior, they must be investigated by means other than the natural sciences.

Finally, having briefly reviewed some of the principal working assumptions of the natural sciences, what might we say concerning the *warrants* for these assumptions? On what basis are these presuppositions made? On mankind's desire for order? On our observations and our experience with natural phenomena? On human creativity and imagination? On our intellectual curiosity to propose and test hypotheses? On the cumulative insight gained from past curiosity, experience, observation, and creative organization of information? If my understanding of the history of science is correct, all of these factors have contributed significantly to our list of working presuppositions.[4] The repeated validation of these assumptions, when they have been applied to numerous phenomena subjected to careful scrutiny, serves as the warrant for the conjecture that they apply to all phenomena in the category of material behavior. While we, as scientists, are not claiming logical proof (for that would be im-

4. For those readers who wish to explore the fascinating world of the history of science, I suggest the following: Charles Coulston Gillispie's *The Edge of Objectivity* (Princeton: Princeton University Press, 1960); Herbert Butterfield's *The Origins of Modern Science,* rev. ed. (New York: The Free Press, 1957); and Richard S. Westfall's *The Construction of Modern Science* (Cambridge: Cambridge University Press, 1971). Each contains an extensive bibliography and suggestions for further reading.

possible), we are demanding that the insights gained by the application of these presuppositions be respected as worthy of serious consideration. Furthermore, because of the wealth of insight into material behavior that has been gained on the foundation of these assumptions, we believe that the first step in taking the material world seriously in the twentieth century is to respect the credibility of the presuppositions on which the scientific study of that world is founded.

THE DOMAIN OF SCIENTIFIC INVESTIGATION

The scientific enterprise has often been accused of either claiming or desiring to be the source of answers for all categories of questions. Now it must be admitted that there are some persons, both inside and outside of the scientific community, who would make the bold claim that all meaningful questions can in principle be answered by means of scientific investigation and that any question not answerable by this method is simply not meaningful. That claim, however, comes not from the disciplines of the natural sciences but rather from the philosophical realm. The claim that scientific inquiry provides the only avenue to knowledge is called *scientism,* not science. Such an exclusivist attitude has no place within the scientific enterprise as such, in spite of what a handful of vocal popularizers of science may say. Good, honest natural science necessarily recognizes the boundaries of the domain in which it can legitimately be applied.

The scope of natural science is limited to the study of the physical properties and behavior of material systems, their interactions with one another, and the character of the space-time continuum in which they function. The sciences deal exclusively with the material world treated in isolation from any nonmaterial influence. Natural science takes a strictly internal approach—that is, it considers only the internal interactions of one part of the corporeal cosmos with other parts of that same cosmos. If there are external, nonmaterial influences or interactions, they are not subject to scientific investigation. Science can neither affirm nor deny anything about them; it must remain silent. The sciences deal only with the "internal affairs" of the cosmos.

It is for this reason that questions about the status of the cosmos, about the transcendent cause for its existence or behavior, about its value, and about its purpose lie outside the domain of the natural sciences. Such questions must necessarily take into account the matter of the relationship of the cosmos to non-material entities or beings, concerning which the sciences as such have nothing to contribute. The study of the corporeal world in relationship to other forms of being must necessarily be assigned to philosophy and theology. Perceptive and honest scientists understand this limitation and demand that the borders of science not be violated; but there are many popularizers of the scientific enterprise who cross the border and thoughtlessly trespass on foreign territory without openly announcing their intentions.

Having noted these general points about the domain of scientific investigation, let us briefly explore some of the territory within that domain. Specifically, let us look at the areas of physical properties, material behavior, cosmic history, and cause-effect relationships.

Matter as a substance (such as hydrogen or helium) and material systems (such as stars, which are made of hydrogen and helium organized into a structured system) have numerous physical properties. The observation and measurement of these perceptible properties (such as mass, density, temperature, chemical composition, size, color, or structure) clearly lie within the domain of natural science. But matter and material systems are not static entities; they dynamically participate in processes of change. Hydrogen atoms within a star, for instance, are in constant rapid motion. Collisions between atoms may strip away their electrons, leaving behind unshielded atomic nuclei. At sufficiently high temperatures, the motion of the nuclei is so rapid that when they collide with one another they fuse, producing a larger nucleus. The energy generated by this fusion process generates the high temperature of a star's surface and causes it to radiate light. This patterned behavior of matter and systems of matter is also the legitimate object of scientific investigation (we will look at it in more detail in Chapter Eight).

While we can readily understand how the perceptible properties and the directly observable behavior of material systems are the proper object of scientific study, many people have difficulty in understanding that some physical properties and material behavior can be legitimately studied even when they are not directly observable. There are, for example, cases in which objects we would like to study can be seen only indirectly, in terms of their effects on other observable objects. The study of the behavior of the sun's interior by its effect on visible surface properties is one example. But that is a rather complex phenomenon, so let us consider a simpler case.

Binary star systems are made up of two stars that both orbit around a point called the system's "center of mass." The orbital motion of each star is determined by the gravitational force exerted on it by the other star. Sometimes both stars of a binary system are bright enough to be observed directly, but in many cases only one will be visible. Even in such cases, however, the presence of the unseen star can be readily discerned by its gravitational effect on the visible one. The presence of the invisible star and at least one of its properties (the value of its mass) can be determined from the observable consequences of its behavior as a member of the binary star system.

We make similar second-hand observations every day. If someone were to ask you if there is a wind blowing outdoors, you wouldn't attempt to observe the movement of air directly; you would look for the effects of the wind on the leaves of the tree outside your window. If you saw the leaves fluttering you could confidently report, "Yes, there's a nice breeze blowing; let's head for the lake and go sailing." The evidence for wind is indirect but convincing nonetheless. Inferences concerning the existence and behavior of one material system by its perceptible effects on another are an accepted component of both natural science and ordinary experience with natural phenomena.

Another circumstance in which indirect evidence must be employed is the study of behavior, processes, and events that occurred some time in the past and are no longer directly observable. While we cannot generally observe a past event directly (with the exception of some fascinating astronomical phenomena that we'll discuss later), we may be able to observe the perceptible consequences of that event—the physical record the event has produced—and from that record we can in many instances infer both the character and chronology of the event or process. A comparison with similar behavior displayed by comparable systems at the present time can then aid us in evaluating our reconstruction of the event. For example, the observed geological structure of the Hawaiian islands and observations of the volcanic activity occurring there now provide the basis for a study of the history of the formation of those islands. Similarly, the properties and behavior that the sun and other members of the solar system presently exhibit provide numerous clues concerning the system's history. The large number of craters observed on the surfaces of the moon, Mercury, Mars, and the satellites of Jupiter and Saturn reveal an extended period of meteor impact on the surfaces of these bodies, for instance.

The essential point here is that the present properties, structures, and behavior of material bodies provide a wealth of information concerning the history of those bodies. The cosmos is permeated with evidence of its own history. The study of that history lies legitimately within the domain of the natural sciences.[5] While the conclusions drawn from such a study are often tentative, and while the reasoning leading to many of these conclusions often takes the form of a plausibility argument, I firmly believe that the conclusions must be given serious consideration. We ought never to give them a casual or thoughtless dismissal; that would be neither good science nor honest dealing with available resources of information.

In the course of studying the material universe, natural scientists ordinarily first seek adequate *descriptions* of physical properties, material behavior, and historical records of events and processes. Having compiled such descriptions, our human curiosity compels us to seek *explanations*. Why these properties? Why this particular behavior? What happened in the past to create this present appearance?

The explanations that natural science is able to offer fall into the category of cause-effect relationships—more specifically, proximate or immanent cause-effect relationships, which, as we noted earlier, are statements of correlation between properties and behavior. Scientific "explanations" cannot go beyond these bounds; they cannot deal with ultimate or transcendent causes, because that might involve the action of nonmaterial entities or beings, which are by their very nature beyond the reach of scientific investigation. When operating within its proper domain, science does not compete with philosophical or religious considerations, because, unlike philosophy or religion, natural science is limited to the study of the corporeal world alone. Thus, scientific "explanations" of phenomena never constitute alternatives to, or competitors with, philosophical or religious statements. They are simply ideas that have to be

5. The realization that cosmic history and chronology could be inferred and computed from a knowledge of properties and processes observed at the present time is a relatively recent phenomenon. For a historical review of this development, see Stephen Toulmin and June Goodfield's *The Discovery of Time* (Chicago: University of Chicago Press, 1965).

satisfactorily incorporated into a much larger philosophical or religious system of thought—a worldview. We will come back to this matter again in Part III of this book; for the moment let it suffice to say that careful attention paid to the definition of the domain of the natural sciences will aid greatly in making the necessary distinction between natural science per se and larger systems of thought (worldviews) into which it has, to varying degrees of success, been integrated.

In sum, the domain of the natural sciences is the investigation of the perceptible properties, observable behavior, and discernible consequences of the behavior of matter and material systems. Natural science may legitimately address questions that not only seek an accurate description of material properties and behavior but also investigate the sequence of proximate or immanent causes of corporeal phenomena. Cosmic history is the proper object of scientific study. The relationship of the cosmos or its history to nonmaterial entities or beings, however, is not a matter for scientific investigation. Questions concerning the status, transcendent cause of existence, transcendent governing power, value, or purpose of the cosmos cannot be answered by the sciences.

To take the material universe seriously, we must know not only the strength of scientific investigation but also its limits—what lies within the domain of natural science and what lies outside of that domain. That's not just good philosophy; that's good science—the kind of science that the vast majority of scientists seek to practice.

THE METHODS OF SCIENTIFIC INVESTIGATION

The elements of scientific methodology fall into two principal categories: empirical and theoretical. The empirical category includes all methods based on observation and measurement. Observation may be as simple as a direct sense perception or as sophisticated as the operation carried out by the information gathering and transmitting system of the Voyager spacecraft, which provided us with those spectacular views of Jupiter, Saturn, and their moons. Empirical science incorporates not only the results of observation and measurement but also the discovery of various relationships among the properties and processes that have been observed.

Theoretical science is a bit more elusive. Because it often employs mathematical manipulation beyond the reach of those

Above, the ringed planet Saturn, as viewed from the Voyager 1 spacecraft when it was still 13 million miles from the planet. This remarkable photograph, like all other Voyager images of Saturn, was constructed from radio signals transmitted over a distance of a billion miles to receiving stations on earth. Also visible on this photograph are three of Saturn's moons—Tethys, Dione, and Rhea. To the left, Enceladus, a 300-mile-diameter satellite of the planet Saturn, as seen by the Voyager 2 spacecraft from a distance of 75,000 miles. Its icy surface contains the record of an active past: parts are pocked by meteor impact craters, much like our own moon, but the large smooth areas suggest episodes of flooding. Note how the circled crater appears to have been cut in half as a result of surface activity.

Photos courtesy NASA and the Jet Propulsion Laboratory.

who haven't been trained to employ it, it is less well understood and, in some instances, less trusted. The basic idea and the principal components of theoretical science aren't all that difficult to grasp, however. One component we've already had occasion to mention is the postulation of proximate cause-effect relationships. We observe an apple fall earthward and, standing on the shoulders of Newton, we postulate that this behavior is caused by the force of gravity exerted by the earth on the apple. In this illustration we have a relatively simple combination of empirical observation and theoretical "explanation." As is typical, the explanation is limited to the expression of immanent cause-effect relationships, or statements of property-behavior correlation.

Theoretical science is also characterized by the closely related practices of generalization and unification. Consider the phenomenon of gravity again. As Newton first discovered, the same kind of explanation that accounts for the falling apple phenomenon also satisfactorily accounts for the moon's orbital motion about the earth. But gravitationally induced orbital motion can be seen on still larger scales. Gravitational theory applies, as we saw earlier, to the orbital motion of planets around the sun and to the orbital motion of binary stars; it even accounts for the motion of billions of stars around the center of our galaxy, and beyond that to the motion of entire galaxies relative to one another. For this reason Newton's description of the gravitational interaction force is often called the "universal law of gravitation."

The particular behavior of falling apples is but one example of a general, or universal, phenomenon. Led by extraordinarily creative insight, Newton generalized from the particular behavior of apples and moons to the interaction of all bodies with mass. This generalization allows us to unify our understanding of systems with vastly different dimensional scales in terms of the concept of gravitational force and its effects on motion. The generalization from particular to universal patterns and the unification of diverse phenomena under a single theoretical framework are especially important and productive components of modern natural science.

A third component of scientific methodology that deserves mention is "theoretical modeling."[6] In a sense, this is the scien-

---

6. For a stimulating discussion of the role of models in both natural science and religion, see Ian G. Barbour's *Myths, Models, and Paradigms* (New York: Harper & Row, 1974).

tific equivalent of the literary device of simile or metaphor. If we wish to investigate (observe, describe, and explain) a very complex phenomenon, we may well find it helpful to compare the behavior of the complex system with the behavior of a simpler, hypothetical system that lends itself more readily to description and explanation. We might compare the behavior of the molecules in a sample of hydrogen gas, for example, to the behavior of a large number of tiny, perfectly hard, rapidly moving spheres that collide frequently with one another and with the walls of a container. That simplified model (a mental construct, not a physical one) works reasonably well to explain why the pressure, volume, and temperature of the gas should be observed to exhibit a particular relationship. But a study of the temperature changes that occur as a result of adding heat to the gas reveals that the model is inadequate. It can be significantly improved, however, by comparing the molecule of hydrogen not to a sphere but to a miniature, stretchy dumbbell: two points of mass separated by a massless spring. This model allows us to see how energy might be stored not only in the motion of the molecule as a whole but also in its rotational and vibrational motion. Even that improved model would have to be modified to account for the fact that at lower temperatures the vibrational or rotational motions may not be activated.

In any case, the details of the kinetic-molecular theory of gases are not of great consequence to our discussion. The point here is that models are an important part of the methodology of theoretical science. It can be very illuminating to analyze the behavior of complex systems by comparing them to simpler conceptual models—models that can be allowed to become progressively more complex so that the model behavior more closely resembles the real system behavior. While the model and the real system can never be equated, as the model's properties and behavior come more and more closely to resemble the actual system's properties and behavior, we judge that the model is a good one and that its comparison to the actual system is worthy of serious consideration.

Within the scientific community there is much good-natured sparring between those individuals engaged primarily in empirical work (observation, measurement, experimentation) and those who occupy themselves mainly with theoretical considerations (computation, modeling). The point of contention is a familiar one: Who are greater in the kingdom of science, the theorists or the experimentalists? The former take special pride in their mathematical prowess, the latter in their laboratory skills; both pride themselves on their analytical acumen and creative insight. Clearly, both make contributions that are vital to the scientific enterprise.

But for the record, let me note that any theory is vulnerable to criticism based on observational testing. If a theory fails the test, it must be modified or discarded. Observation rules over theory. That says nothing about the relative importance of the theorist and the experimenter, but it does say something about the methodology of the natural sciences. Science is not free to create fantasy; it is bound by the constraints imposed by the real world. It is *this* world that we seek to understand the properties, behavior, and history of; make-believe worlds are of little interest to science.

A final aspect of scientific methodology that warrants discussion is its public character. By "public character" I mean that the data, analysis, and conclusions of scientific investigation are open to public scrutiny. And scrutiny they get! Sloppy or dishonest work in the sciences is not tolerated. There have been short-lived exceptions to this rule, but in the long run accuracy and honesty are demanded. It is a common occurrence for one scientist to expose the lack of precision in measurements or the inadequacy of experimental procedure or the faults in the reasoning employed in the published work of another scientist. I suspect that this is a natural expression of the human ego; to expose another's faults appears to strengthen one's own position. In any case, this public dimension of scientific work provides a mechanism for self-correction in natural science. That doesn't make the scientific enterprise infallible, but it certainly discourages runaway error and provides us with further confidence that natural science can serve as a legitimate means of taking the material world seriously.

THE RESULTS OF SCIENTIFIC INVESTIGATION

What can we say about the trustworthiness of the published results of empirical or theoretical natural science? How much credibility do scientific descriptions, explanations, and theoretical models deserve? Must we accept everything scientists publish as the truth, the whole truth, and nothing but the truth?

There is no doubt in my mind that the results of scientific investigation deserve serious consideration; indeed, I firmly believe that such results are on the whole entirely credible. On the other hand, neither I nor any other scientist would claim that they are infallible. Science is an enterprise carried out by ordinary human beings with finite abilities and knowledge—highly skilled and knowledgeable in certain areas, but capable of making mistakes.

However, the admission of fallibility on the part of scientists provides no warrant whatsoever for the wholesale dismissal of the results of scientific investigation or the cavalier rejection of any major explanatory framework. Modern natural science stands as a paragon of analytical achievement in the effort to understand certain aspects of the world in which we live and of which we are a part. The scientific enterprise ought not to be the object of unquestioning trust, but it has rightly earned a high degree of respect.

Beyond this, specific results of scientific investigation deserve credibility in proportion to the weight of the supportive or corroborative evidence on which they are based. That particular description, explanation, or historical reconstruction that makes best use of the greatest amount of evidence deserves the highest credibility. Scientific explanations seek to relate causally a particular phenomenon or event to its spatial, temporal, and material environment. Any explanation that does that well is worthy of further consideration.

Specific explanations or theories deserve credibility in proportion to their coherence and integrity relative to the whole array of data and other explanatory structures. If you were offered two theories, one based on a vast collection of consistently related phenomena and the other based on a small number of isolated anomalies, which one would you find more credible? The natural scientist would invariably choose the one based on the larger body of coherent evidence. That's just common sense.

Or suppose that you were offered two explanations of some observation, one that is coherent with the most general principles applicable to the patterned behavior of matter and another requiring a violation of these same firmly established principles. Obviously, the explanation that is coherently related to the whole spectrum of accepted statements concerning the patterned behavior of matter is the more credible. Before we could consider the other choice credible, we would have to show that it not only explains the phenomenon in question best but also that all other relevant phenomena can be more satisfactorily interpreted or explained in the new way. The burden of proof for such major changes in paradigm is overwhelming. Theories or models that demand changes of this magnitude invite the most meticulous scrutiny. Few survive.[7]

Finally, suppose that two theories are proposed, one based on limited but directly accessible data, and another based principally on the absence of contradictory evidence. Which of these two is to be preferred? As a general rule, arguments from silence are weak, certainly weaker than arguments from positive evidence, and scientific theorizing is no exception to this general rule.

How credible is a scientific theory or the result of a scientific investigation? It should be assigned a credibility in proportion to the care and impartiality with which observations were made, in proportion to the degree to which the evidence positively supports the conclusion, and in proportion to the degree to which it is coherently related to the entire body of understanding concerning material phenomena. Furthermore, the result of a scientific study should be assigned a credibility in proportion to the extent to which it has made full use of the legitimate methods of scientific investigation and in proportion to the degree to which it has honored the limitations of natural science and stayed within its legitimate domain.

Blind, unquestioning acceptance of the results of scientific investigation is neither appropriate nor desirable. At the same time, wholesale denial of strongly supported interpretative frameworks, and the substitution of incoherent ad hoc interpretations that fly in the face of accepted explanations of the cosmos strike me as equally inappropriate and undesirable. We will take a look at some examples of that sort of nonsensical explanation in Chapter Eleven.

A CONTEMPORARY VIEW OF THE COSMOS

When we look at the material world in the manner of the natural scientist, what does it look like? What general properties does it exhibit? What kind of behavior does it display? What kind of history has it experienced? Obviously we can't supply complete answers to such large questions, but I would like to take a look

---

7. Revolutions in conceptual frameworks occur when the weight of unresolved anomalies becomes greater than the old framework can bear and when a better framework is available. For the landmark analysis and discussion of this phenomenon, see Thomas S. Kuhn's *The Structure of Scientific Revolutions*, 2d ed. (Chicago: University of Chicago Press, 1970).

at a relatively small number of important general features of the properties, behavior, and history of the cosmos that will be important for our later discussion of the nature of stars.

Some General Properties Exhibited by the Cosmos

Throughout most of human history it was thought that the cosmos was divided into several "realms" that differed from one another in numerous ways. The Hebrews, along with other ancient Eastern cultures, often spoke of a threefold subdivision of the cosmos into the heavens, the earth, and the underworld. Early Greek pictures of the cosmos were based on a twofold division: the celestial realm and the terrestrial (or sublunar) realm.

Modern natural science, however, has discovered that the cosmos is neither a "triverse" nor a "biverse" but a "universe" characterized by a oneness of both substance and behavior. The whole cosmos is made up of the same elementary particles, chemical elements, and energetic radiation. The sun and other stars are made mostly from ordinary hydrogen and helium. A distant galaxy is built from the same chemical elements that our own bodies are made of. The whole cosmos is constructed of the same elements that we find right here in the dust of the earth.

Though the cosmos is universal in substance, there is a hierarchy in its structures. Atoms and alligators do not differ in the kind of substance of which they are made (neutrons, protons, and electrons), but they certainly differ in the complexity of their structure. Atoms are built of certain elementary particles. Molecules are built of atoms. The cells of a living organism are intricate structural arrangements of highly complex molecules. The body of a living creature (such as an alligator) represents an astounding structure of specialized cells functioning in a marvelously organized manner. Similarly, a star represents a structure built of atoms. Hundreds of billions of stars arranged in a spiral-structured disk constitute a galaxy. An equally large number of galaxies form the largest structure known—the visible universe. Thus the universe we observe and study appears to be made of a small number of fundamental types of units assembled into a hierarchy of progressively more complex structures extending from the subatomic to the extragalactic in dimensional scale.

Closely related to the great differences in the complexity of corporeal structures is the vastness in the dimensional scale of these structures. Toward the lower end of this scale we find the atomic nucleus, which is so small that along a line one inch long more than a million million nuclei could be placed side by side. An atom is about ten thousand times larger, but is still small relative to the objects that we perceive in our daily experience. A star, by contrast, is proportionately larger than the things we ordinarily encounter. Compared with the size of a human being, an atom is billions of times smaller, while a typical star is billions of times larger. Yet on the astronomical scale even stellar dimensions are miniscule, dwarfed, for instance, by the distances between stars—typically a few light-years.[8] Galaxies have dimensions on the order of 100,000 light-years and are separated from their neighbors by distances of many millions of light-years. The most distant objects known are located billions of light-years away, and according to current cosmological theory even this distance is small relative to the scale of the spatial framework within which the visible universe functions. Yet, while the scale of cosmic dimensions is incomprehensibly vast, most contemporary models of the cosmos suggest that it is finite. The cosmos is indeed vast, but it is not limitless.

Some Aspects of Material Behavior

We need note little more about material behavior than we have already noted in our discussion of the assumptions of natural science, which are distillations of our experience with the material world and form the basis of the contemporary view of the cosmos. Natural science assumes the behavior of matter and of material systems to be rigorously patterned, not chaotic or capricious. We typically call the statements describing those patterns of behavior the "laws of nature." Furthermore, we believe the patterns of material behavior to be spatially universal and temporally invariant: they are the same everywhere and "everywhen." This uniformity of behavior extends not only over space and time but also over the levels of structural complexity found in nature. Atomic nuclei as well as clusters of galaxies, for example, exhibit behavior that satisfies the principle of energy conservation. Finally, the patterned behavior of

8. The "light-year" is a unit of distance defined to be the distance that light travels (at a speed of 186,000 miles per second) during one year. It is equivalent to about 6,000,000,000,000 miles, or approximately 1,500 times the distance from earth to the planet Pluto.

M 31, the Andromeda galaxy, located at a distance of two million light-years, is the nearest major spiral galaxy in our neighborhood. The light that we are now receiving from this luminous giant began its journey to earth two million years ago.

matter is related to material properties in ways that can be described by immanent cause-effect relationships. What matter *does* is related to what matter *is*. Behavior follows from properties according to definite rules.

As far as we can tell from thorough scientific investigation, the behavior of the cosmos follows a coherent set of spatially uniform and temporally invariant patterns that are causally (in the proximate sense) related to material properties.

Cosmic Chronology

Because material properties are universal and because material behavior, which is causally related to these properties, is coherent and invariable, the natural sciences can legitimately study cosmic history. And I believe that cosmic chronology can be reliably inferred from the physical record formed by events and processes in cosmic history. This chronology has a number of fascinating features that we should look at briefly at this point.

Cosmic chronology is revealed in a large number of independent phenomena. The phenomena themselves are numerous and diverse, as are the ways we have of determining the times at which they occurred, and yet altogether they paint a remarkably coherent portrait of cosmic history. The cosmic timetables suggested by physical cosmology, stellar astronomy, nuclear physics, geophysics, geology, chemistry, and biology are all essentially the same.

According to evidence gathered through each of these disciplines and others, cosmic history is made up of a coherent sequence of causally related processes and events spanning several billions of years. (We will be looking at selected examples of this evidence in Chapters Eight and Nine.) This time scale staggers the imagination and certainly dwarfs the human experience of time. But although fifteen billion years is a long period of time, it is—very significantly—finite. Cosmic history can be traced only as far back as the "big bang," that remarkable event that signals the beginning of the universe we observe. As I have already suggested, the natural sciences cannot determine the meaning, significance, or ultimate cause of such an event (or of any other event, for that matter), but they can legitimately investigate the physical character of that event and the time at which it occurred.

The appearance of the sun, the earth, and earth's geological structures, along with the plants and animals that populate the earth's surface are parts of cosmic history that are coherently related to all that has happened—events at all places and times. Terrestrial history cannot be isolated from cosmic history. Whatever agent is governing the course of history in the universe is governing it in such a way that events occurring at one time or place are coherently related to all other events, regardless of when or where they occur, so that there appears to be a directionality to history. Cosmic history doesn't merely happen; it is going somewhere. The discovery of this phenomenon by scientific investigation raises many intriguing questions about transcendent cause and purpose, but again these are matters beyond the domain of the natural sciences. To answer the profound questions that humanity is naturally inclined to ask requires more than natural science can offer.

CHAPTER SEVEN

The Scientific Study of Stellar Properties

STARS AND STARLIGHT

Because the stars are so incredibly distant from us, you may be wondering how we can learn anything about them—how we even know how far away from us they are. The answer is that we learn about stars from starlight. All information concerning the physical properties of a luminous celestial body such as a star is carried to us by the light and other related forms of radiation it generates. The amount of information we can accumulate about the stars depends on our ingenuity in devising ways of analyzing starlight.

If we are to gain some appreciation for the validity and integrity of what astronomical investigation has revealed about stars and a host of other celestial bodies, we must first acquaint ourselves with the character of the data, the analytical procedures, and the results of that scientific study. In particular, we must learn something about the nature of light itself and how the properties of a source of light affect the measurable properties of the light it produces. Our goal, then, is the same as that of contemporary astronomy: to learn how to analyze the light we receive from some remote celestial body in such a way as to determine as many of the properties of that body as possible. This isn't a textbook on astronomy, though, so we won't try to be exhaustive; for a general understanding it isn't necessary to be highly technical anyway, because as it turns out the scientific investigation of stars has a lot in common with ordinary experiences here on earth.

As I write this paragraph, I am privileged to be sitting by a window overlooking Lake Michigan. With considerable confidence and certainty, I would claim that there is now a sailboat (a sloop sailing on a broad reach under genoa and mainsail) passing by. Why do I make this claim with confidence? Because there is light coming from the lake and the sailboat, conveying that information to my eyes, and because my brain is able to interpret the signals it is receiving from the optic nerve in such a way as to inform me of this beautiful sight. (Another part of my brain recognizes that I would enjoy being on that boat, sailing under blue skies in a moderate northwest breeze, but that's another story.) In any case, the fact that I can believe my eyes when they tell me about a boat on a lake suggests that I can also believe my eyes (and other specialized devices designed to gather and process light) when they tell me about the stars.

INFORMATION IN THE ENERGY CONTENT OF STARLIGHT

Light is a form of energy. Starlight is a stream of energy running from a star to an observer. The energy content of starlight conveys information about some of the star's physical properties. To find out more about how it does this, we have to know more about energy in general.

Energy is often defined to be the capacity to do work. That's fine, provided we have an adequate definition of "work." For the purposes of this discussion, let's just say that work is done when a force acts on an object and causes it to move or to change its state of motion. The quantity of work done is related to the product of the force and the distance through which the object was moved, or displaced.

Energy, then, is the capacity to exert force and cause displacement. It comes in a variety of forms. "Kinetic energy" is the energy an object has by virtue of its motion. A moving car possesses kinetic energy, for example; it would be foolish to stand in the path of a moving car, because we know that it has the capacity to exert force and to cause displacement. "Potential energy" is the energy an object has by virtue of its position relative to another body. If I were to hold a large stone ten feet directly above your head, you would probably be uneasy—and rightly so, because you know that that stone possesses gravitational potential energy: by virtue of its position relative to the earth, it has the capacity to exert force and to cause displacement. And displacement of your head can ruin your entire day!

In addition to kinetic and potential energy, other common forms of energy include heat energy, chemical energy, electrical energy, nuclear energy, and light energy. The energy content of sunlight, for example, is vital to the maintenance of life on earth and may become an increasingly important source of energy for conversion into the heat or electricity required to maintain comfortable and convenient homes.

Conversion of energy from one form to another is the most basic of physical processes. Living plants convert sunlight into various forms of chemical energy. Plant products, such as oil and coal, are burned to produce heat, and the heat can be used to form steam to drive turbines that in turn drive generators to produce electricity. Electrical energy is conveyed by wires into our homes to generate heat and light and to power appliances that make countless tasks safer and easier.

While virtually every physical process involves the conversion of energy from one form to another, the quantity of energy remains constant. We have never observed a process in which the total amount of energy has changed; energy can neither be created nor destroyed by any known physical process. This principle, the law of energy conservation, is one of the most fundamental statements describing the manner in which material behavior is governed. We will have frequent occasion to apply it in future discussions of stellar properties, behavior, and history.

To specify the quantity of energy in a system or the amount of energy transformed from one type to another, we must have some convenient units. Names of familiar energy units include the "calorie" and the "BTU" (British thermal unit). A unit that is less familiar to most people but that will be more convenient for the present discussion is the "joule." One joule is the amount of energy needed to lift one pound a distance of nine inches. It takes about 70 joules of heat energy to raise the temperature of one ounce of water 1°F. Burning one gallon of gasoline would convert approximately 100 million joules of chemical energy into heat. A 60 watt light bulb consumes electrical energy at the rate of 60 joules per second. (One watt is defined to be one joule per second.) If the light bulb were 100 percent efficient, it would produce light energy at the rate of 60 joules per second, but in ordinary light bulbs part of that energy is converted into heat.

Now, what has all of this energy talk to do with stars? Stars are luminous objects, which is to say that they generate light energy. The rate at which a star generates light energy is called its "luminosity." A star's luminosity value can be specified in the same way as that of a light bulb: in joules per second, or watts. The luminosity of the sun, for example, has the value of 4×10^{26} watts.[1]

The sun's luminosity is so great compared with the luminosity of familiar terrestrial light sources that it is difficult to comprehend. Perhaps an illustration would help. Suppose there were a light bulb factory that produced 100 watt bulbs at the rate of 1000 per second (no backyard operation!). How long would it take that factory to produce enough bulbs so that when all were turned on they would together equal the luminosity of the sun? A year? A thousand years? A million years, perhaps? A simple computation will show that the time required for this (rather unrealistic) task is more than a hundred thousand billion years! A fruitless effort, of course. Turning on all of those bulbs would probably produce not light but merely a blown fuse.

When we look at the stars of the night sky, some are obviously brighter than others. Everyday experience suggests that such differences in apparent brightness could be the result of either of two causes or both. One star may appear brighter than another because it is more luminous or because it is closer (as a streetlight appears brighter than the brightest of stars in the night sky because it is closer). There is, in fact, a very definite mathematical relationship among the values of brightness, luminosity, and distance for light sources like stars. However, in order to state that relationship we must specify more precisely what we mean by the term *brightness*. Quantitatively, we define brightness in terms of the rate at which an observer is receiving light energy in a specified amount of light-collecting area. It is the same type of quantity that is measured by a photographer's light meter. For our purposes, it will be convenient to express this value in watts per square meter. The sun, the brightest star in our sky because of our proximity to it, has a brightness of approximately 1,000 watts per square meter as viewed from the earth. For the sun, or a star, or any other celestial light source that radiates uniformly in all directions, the value of its apparent brightness can be calculated using the following equation:

---

1. 4×10^{26} is a convenient shorthand notation for the number 4 followed by 26 zeros.

$$B = L/4\pi D^2,$$

where:

B is the star's brightness in watts/meter^2,
L is the star's luminosity in watts, and
D is the star's distance in meters

Two applications of this brightness-luminosity-distance relationship are particularly useful to astronomers. First, if both the distance and brightness of a star can be measured directly, then its luminosity can be computed from this relationship. This is the case for several thousand of the nearest stars the distances of which can be determined by direct geometrical means.[2] The second application is the computation of the distance to stars for which we know the brightness and luminosity values.

But that raises a question, doesn't it? How do we come to know the value of a star's luminosity if we don't yet know its distance? If I were to give you a light bulb and ask you to tell me its luminosity, the simplest way for you to determine it would be to read the label printed on the bulb. A very reasonable procedure for light bulbs, but will it work for stars? As it turns out, many stars do have labels that reveal their luminosity values—not written in words or numerals, of course, but encoded in other recognizable features.

From the nearest stars, the luminosity values of which can be computed from measured values of brightness and distance, astronomers have learned to recognize definite correlations between certain stars' luminosity values and other characteristic properties revealed by their light. The features of these special stars allow us to estimate their luminosity with some precision, and with that information in hand, we can use the brightness-luminosity-distance relationship to compute the distance to these stars.

Is there any more information that can be squeezed out of the energy content of starlight? Indeed there is. We can, for example, learn the size of a star if we are able to determine its

2. When we move our head from side to side, we notice that nearby objects appear to change their angular position relative to more distant objects. This phenomenon is called "parallax." As the earth revolves around the sun each year, we observe stars from differing vantage points. Nearby stars, therefore, exhibit a parallax effect in an annual cycle. By carefully measuring the annual change in the angular position of nearby stars relative to very distant background objects, astronomers are able to compute the distance to these stars.

luminosity and its surface temperature. The Stefan-Boltzmann law, which can be written in the form of an algebraic equation relating the values of a star's diameter, its luminosity, and its surface temperature, can be used to supply this value. In fact, this is practically the only means astronomers have to determine stellar sizes; only for the sun and a handful of nearby large stars can we measure sizes by geometrical techniques.

But there's a catch, isn't there? In order to compute a star's size using the Stefan-Boltzmann law, we have to know not only its luminosity but its surface temperature. How can this temperature value be determined? Does starlight reveal this stellar property also? To answer this question we must discuss more of the fundamental properties of light and their relationship to the temperature of a heated surface that generates it.

INFORMATION IN THE WAVELENGTH SPECTRUM OF STARLIGHT

The Electromagnetic Spectrum

Visible light is just part of a whole band of radiant energy phenomena that also includes gamma rays, X rays, ultraviolet light, infrared radiation, microwaves, and radio waves. All of these phenomena are called "electromagnetic waves," and together they constitute what we call the "electromagnetic spectrum."

What is an electromagnetic wave? Perhaps the best way to give meaning to this technical term is first to consider some more familiar wave phenomena. A water wave, for example, may be conveniently described as a traveling variation in the height of a water surface. Similarly, a sound wave might be described as a traveling variation in air pressure. Following this same pattern, then, we could describe an electromagnetic wave as a traveling variation in electric and magnetic field strengths.

Fine. Waves are traveling variations in something. For water waves and sound waves the something is a material property that can readily be visualized—water surface height or air pressure. However, for electromagnetic waves the something is not so familiar. What on earth are electric and magnetic fields? For our purposes we needn't get technical about it (though it is certainly possible to do so—a college physics major program includes about three semester courses on that subject); it will be enough to say that electric and magnetic fields are modifications in the properties of space that will exert a force on certain objects. An electric field, for instance, will exert a force on any object that is electrically charged, such as strands of combed hair on a dry winter morning. A magnetic field will exert a force on objects such as a compass needle. A compass needle points to the north because the earth naturally generates a magnetic field that exerts a force on it—in this case, a force causing it to align with the north and south poles.

The strength and direction of electric and magnetic fields can vary from point to point. If the source of the field changes, the ensuing change in the field itself will travel away from the source at a characteristic speed, commonly called the "speed of light." Light and all other electromagnetic waves will travel through empty space at the dazzling speed of 186,000 miles per second—more than seven times around the earth's equator in a single second. The simplest pattern of field strength variation that travels at this astonishing speed is illustrated in figure 7-1. It is a smoothly varying, "wavy" pattern that repeats at regular intervals.

Electromagnetic waves can be characterized by the value of their *wavelength,* the distance between successive equivalent points on a wave train. Visible light has wavelength values in the range of 4000-7000 Angstroms.[3] Within that range, our eyes perceive different wavelength values as different colors. The rainbow provides a good illustration of the wavelength-color relationship. Raindrops act on sunlight in such a way as to disperse sunlight, which is made up of a mixture, or spectrum, of many differing wavelength values, into its constituent parts. The bands of color we observe are arranged in the order of their corresponding wavelength—from violet (the shortest visible wavelength) through blue, green, yellow, and orange to red (the longest visible wavelength). And, as I have already suggested, visible light itself is only part of the larger spectrum of all electromagnetic radiation. From shortest to longest wavelength, this spectrum runs from gamma rays to X rays, ultraviolet light, visible light, infrared radiation, microwaves, and radio waves.

The spectroscope is one of the most important instruments in observational astronomy. Its function is to take light from a given source and break it up into its constituent parts, each with

3. An Angstrom is a unit of length defined to be one one-hundred-millionth of a centimeter; there are 254 million Angstroms to the inch. The wavelength of visible light is about one fifty-thousandth of an inch.

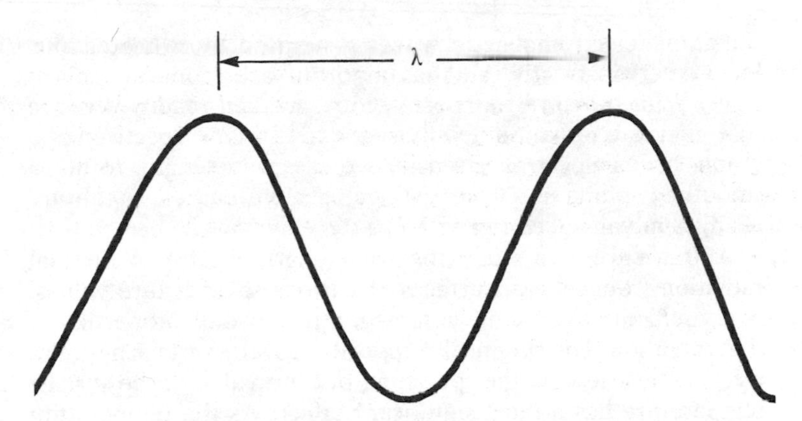

FIGURE 7-1. SIMPLE ELECTROMAGNETIC WAVE PATTERN. The distance from the crest of one wave to the next, here marked by the Greek letter λ (lambda), is called its *wavelength*. The principal difference between various kinds of electromagnetic radiation—between visible light and gamma rays, for example—is simply a difference in wavelength.

different wavelength values, much as raindrops break up sunlight into the colors of the rainbow. One convenient way to present the results of this dispersion of light—the "wavelength spectrum"—is to construct a graph charting spectral intensity against wavelength (i.e., a graph showing the energy content of the light at each possible value for wavelength). Such a graph contains a wealth of information about the source of the light being studied. If the source of light is a surface heated to a uniform temperature, the wavelength spectrum of that light will reveal that fact and it will also yield the value of the temperature of the source—even if it is the surface of a star located millions of light-years away.

Taking a Star's Temperature

Any heated surface emits electromagnetic radiation. At moderate temperatures a surface will likely emit the bulk of that radiation in the form of infrared waves, which we perceive as radiant heat. At higher temperatures, the surface will begin to emit visible light. The heating element of an electric heater or an electric stove becomes "red hot." The common incandescent light bulb generates its light with a metallic filament that is electrically heated to a very high temperature.

An analysis of the wavelength spectrum of such thermal

radiation (electromagnetic waves generated by a heated surface) is very instructive and has important astronomical applications. With the aid of a spectroscope, we can readily observe that thermal radiation exhibits a "continuous spectrum": a graph of its spectral intensity versus wavelength forms a smooth, continuous curve with no discontinuities, or abrupt changes in value (see figure 7-2).

If we were to study the wavelength spectra of thermal radiation from several surfaces at various temperature values, we would discover some interesting and useful properties of that radiation. The chemical composition of the surface has relatively little effect on the spectrum, but the value of the surface temperature has a most significant effect. As the temperature changes, the Planck curve keeps its general shape, but the peak of the curve (λ_{peak} in figure 7-2) moves to a different wave-

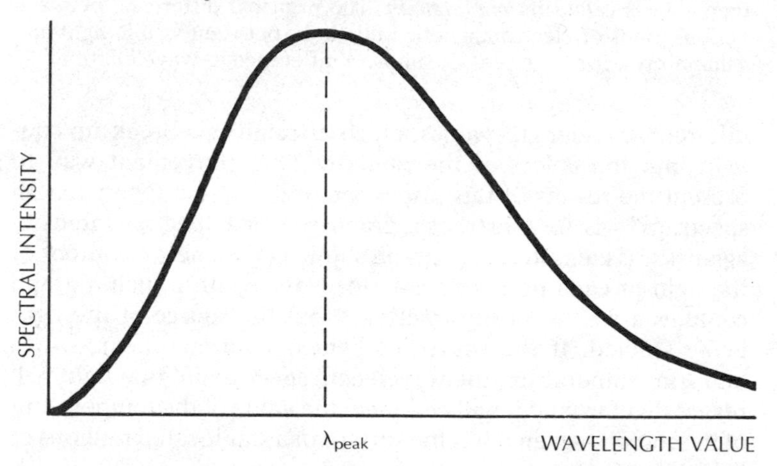

FIGURE 7-2. A REPRESENTATIVE GRAPH OF THE WAVELENGTH SPECTRUM OF THERMAL RADIATION. This particular graph is called the "Planck curve," after the physicist Max Planck. It shows that the spectral intensity is negligibly small at very small wavelength values—a fortunate circumstance for us: it means among other things that we don't have to worry about being showered with heavy doses of X rays every time we turn on an electric light bulb. It is similarly fortunate that the spectral intensity is negligibly small at large values of the wavelength; it means that an ordinary incandescent lamp will generate relatively little energy in the form of radio waves that might interfere with radio or television reception. Very important for our discussion is the fact that the Planck curve displays a single maximum: there is only one wavelength value (marked λ_{peak} on the graph) at which the spectral intensity achieves its peak value for thermal radiation.

length. The manner in which the value of λ_{peak} depends on temperature is remarkably simple; it can be stated algebraically as follows:

$$\lambda_{peak} = 2.9 \times 10^7/T$$

where:

λ_{peak} is expressed in Angstroms, and
T is the surface temperature expressed in degrees Kelvin[4]

In other words, for every given surface temperature there is a specific corresponding value for λ_{peak}. This relationship is called "Wien's law." But Wien's law can obviously be rewritten in a form that is more useful to astronomers: with a simple algebraic manipulation, we can write

$$T = 2.9 \times 10^7/\lambda_{peak}$$

So, for any source producing thermal radiation, we need only measure its wavelength spectrum to determine the value of its λ_{peak}, plug that value into the equation representing Wien's law, and compute the temperature value. This will be the surface temperature of the source we are studying, whether that source is the filament of an ordinary light bulb in the laboratory or the surface of the star Betelgeuse located five hundred light-years away. By analyzing sunlight, for example, we learn that the sun's λ_{peak} = 5000 Angstroms; Wien's law then yields the value for the solar surface temperature: 5800° Kelvin. Once again we see that starlight conveys information about the properties of stars—information that can be straightforwardly extracted by appropriate analytical procedures.

Before leaving this subject of thermal radiation, we should briefly note a familiar consequence of the effect that temperature has on the wavelength spectrum. If you have ever watched an electric stove element heat up, you may have noticed that it changes in both luminosity and color. The change in luminosity is described by the Stefan-Boltzmann law; the color change is related to Wien's law. Although the full explanation of this color change is rather complex, the basic phenomenon can be summa-

4. For scientific purposes it is convenient to work with a temperature scale that starts at "absolute zero"—the temperature at which the heat energy of any object attains its minimum value. The Kelvin scale does so. A Kelvin temperature value is simply the Celsius value plus 273°: O°K is absolute zero, 273°K is the freezing point of water, 373°K is the boiling point of water, and so on.

rized briefly. The temperature value determines the value of λ_{peak}, which in turn affects the relative proportions of the different colors in the visible light being produced by the source. The lower the temperature of the source, the more reddish its color will be. The star Betelgeuse in Orion, for instance, has a cooler surface temperature than the sun and so appears somewhat reddish in color. The bright star Vega in Lyra, which is much hotter, appears white. Stars with still higher surface temperatures appear to be tinted blue. Color photographs of stars readily reveal these color differences. With just a bit of effort, you can even detect some color differences among stars with the naked eye on a clear evening.

What Is a Star Made Of?

So far we have discussed means of determining a star's luminosity, distance, size, and surface temperature. Now we would do well to ask what a star is made of, what it is composed of chemically. Once again, the answer can be found in starlight. Before we can understand how the light conveys this information, however, we'll have to cover some more general information about light and light sources.

The whole material universe is built up from less than a hundred naturally occurring elements, such as hydrogen, oxygen, carbon, iron, uranium, and so on. The smallest unit of an element is the atom, a structure composed of a nucleus surrounded by a system of electrons in motion. The atomic nucleus is positively charged and contains the bulk of the atom's mass. The electrons, which are negatively charged and possess considerably less mass than the nucleus, move swiftly around the nucleus in patterns variously described as clouds and shells and orbits. As a consequence of both their position and their motion relative to the nucleus, electrons possess energy. However, physicists at the beginning of the twentieth century found much to their surprise that the sum of the potential and kinetic energy of atomic electrons is restricted to certain discrete values—a different set of values for each element. An electron must be in some specific state of motion; each possible state involves the contribution of a definite amount of energy to the atom. Thus, the energy an atom possesses by virtue of the motion of its electrons in these restricted states of motion can assume only certain discrete values; no other values are allowed.

Under the proper circumstances, atoms can make transitions from one energy state to another. However, since energy must be conserved, if an atom makes a transition to a state of lower energy, it will have to transfer the excess energy to its environment in some definite amount and form. One such form is electromagnetic radiation: atoms emit light when they undergo a transition from a higher to a lower energy state (by a process different from that which leads to thermal radiation). The energy lost by the atom is carried away by a quantum of light energy called a *photon.* The wavelength of the light associated with that photon is fixed by the amount of energy it possesses—the same amount of energy lost by the atom—and since the energy values of the atomic states involved in the transition are restricted to a set of discrete values characteristic of the type of element to which the atom belongs, each element will emit light having only certain wavelength values. Hydrogen atoms will emit light with a set of wavelengths different from those of the light emitted by carbon atoms, and so on; each of the elements is unique in this way.

The spectrum of wavelength values of the light produced in this fashion is called an "emission-line" spectrum, because when viewing such a spectrum in a spectroscope, one sees a set of bright, narrow lines separated by darkness—each line having a color corresponding to its wavelength value and appearing at a different position in the field of view. This emission-line spectrum functions like a person's fingerprint: the pattern identifies the source. The pattern of lines in an emission-line spectrum reveals the identity of the elements present in the source of light. Many interstellar nebulae—large clouds of glowing gas found among the stars—emit light by the process we have just described. An analysis of their emission-line spectra reveals their chemical composition: mostly hydrogen, some helium, and small amounts of several other familiar elements.

What about stars? What type of wavelength spectrum does starlight exhibit? Stars generate most of their light as thermal radiation from a relatively thin layer of hot gases that we can call the star's "surface." Above that luminous surface, however, lies an atmosphere of somewhat cooler gases the atoms of which can absorb a portion of the surface thermal radiation. This absorption process is precisely the inverse of the emission process: when an atom absorbs a photon, it moves from a lower to a higher energy value. But the energy conservation requirement still applies. An atom must absorb precisely the correct amount of energy to jump from one specific energy state to another specific higher energy level. Since electromagnetic energy comes in the form of photons the energy content of which is fixed by their wavelength value, only light of certain wavelength values can be absorbed by an atom. In fact, because we are once again talking about transitions among atomic energy levels, the wavelength values of light that an atom is capable of absorbing are the same wavelength values at which the atom is permitted to emit light.

What does this mean for the wavelength spectrum of starlight? It means that some of the thermal radiation emitted at a star's surface is absorbed by atoms in the star's atmosphere before we even have a chance to see it. The starlight that we do receive is what is left over—the original thermal radiation from the surface minus the light absorbed at specific wavelength values by its atmosphere. Such starlight exhibits what we call an "absorption-line" spectrum. Starlight viewed through a spectroscope would display the rainbow of colors formed by the continuous spectrum of thermal radiation punctuated by dark absorption lines introduced by the absorption process. The dark lines show specific wavelength values at which light has been attenuated or removed from the total amount of light being generated by the star.

The pattern of absorption lines in a stellar absorption spectrum also represents a type of "fingerprint." It reveals the identity of the chemical elements present in a star's atmosphere. Even more importantly, it reveals the chemical composition of the star itself. Unlike the atmosphere of a planet, the atmosphere of a star is being continuously mixed with the bulk of the star, which is also gaseous. Consequently, the chemical composition of a star and its atmosphere are virtually identical, and so the absorption-line spectrum of starlight reveals the chemical composition not only of the star's atmosphere but also of the entire star.

Analysis of stellar absorption spectra, thermal radiation spectra modified by atmospheric absorption lines, reveals that the vast majority of stars are composed mostly of hydrogen and helium. For a typical star, such as the sun, over ninety percent of the star's atoms are hydrogen; most of the remainder are helium atoms, and fewer than one percent of a star's atoms are contributed by all of the other naturally occurring chemical elements. Nearly all stars are giant luminous spheres of heated hydrogen and helium.

A STAR IS NOT JUST A STAR

Picking a Project

Starlight conveys information about the properties of stars. The energy content and wavelength spectrum of starlight provides information about a star's luminosity, distance, size, surface temperature, and chemical composition. In addition to these physical properties that we have discussed at some length, several other stellar properties are revealed by a star's radiation.

Suppose, then, that we are properly equipped to extract this information from starlight. What would we discover about stars? What are stars like? How do stars compare with one another? Are they all alike? If not, in what ways do they differ?

The project of collecting information about stars to see how they compare with one another may appear to be very straightforward and rather routine, but the method of investigation must be thoughtfully selected, and the results of this research have led to several fascinating surprises. A particularly important peculiarity that has been discovered in the twentieth century is hidden within the rather ambiguous statement "A star is not just a star."

Selecting a Method of Investigation

How shall we carry out our project of comparing stars? What method of investigation might be both convenient and productive? One obvious reply to this question would be to suggest that we simply gather as much information about stars as we can, and then compare the results. While this direct approach might sound good initially, we would soon encounter a very frustrating difficulty: too much information. It would be like attempting to study a forest by trying first to gather every possible item of information about every single tree, branch, and leaf in it. Obviously our first step must be to restrict the scope of our investigation by selecting only part of the available information in order first to discern the larger, overall picture. After the general features have been discovered, the details can be more conveniently filled in.

As a practical matter, we will have to restrict the scope of our investigation both quantitatively and qualitatively. Quantitatively, we will have to restrict the number of stars we consider; we need not study all stars, just a representative sample. Qualitatively, we need not consider all stellar properties, but only those that are necessary to reveal the characteristic differences among stars.

Restricting the number of stars we compare is a straightforward matter, but how do we restrict the list of stellar properties to be considered? The choice is not at all obvious. Early in the twentieth century, two astronomers, Ejnar Hertzsprung and Henry Norris Russell, independently discovered a particularly fruitful choice for the stellar properties to be used as a basis for comparison. Essentially, they chose to compare stars on the basis of only two properties—surface temperature and luminosity.[5]

Having made that choice, our questions of stellar comparison can be stated more precisely: How do stars compare with one another on the basis of their luminosity and surface temperature values? Are the values of surface temperature (T) and luminosity (L) for a given star related? Do stars have preferred combinations of L and T values, or is there no correlation between these properties? Are there different types of stars? Do stars belong to specific families or categories on the basis of their L and T values? Only observation will reveal the answer to these questions, so let's look at some of the results of empirical investigation.

The Results of Observation

Suppose that we have measured the values of the surface temperature and the luminosity of several stars. How do we go about performing a comparison of the results? Merely making a list of the numerical values may be accurate, but it won't readily reveal relationships or patterns in the combinations of values. Hertzsprung and Russell worked out a better method that involves constructing a graph, or diagram, on which the values of L and T for each of several stars are plotted. In fact, such diagrams have come to be called "H-R diagrams" to honor the two astronomers for their important discovery.

It is the H-R diagram that most directly reveals that a star is not just a star. Look at the representative H-R diagram in figure

---

5. Actually they used the properties called "absolute magnitude," and "spectral type." However, absolute magnitude is determined solely by luminosity, and spectral type is determined by the temperature. For simplicity's sake, I'll refer to luminosity and temperature in this discussion.

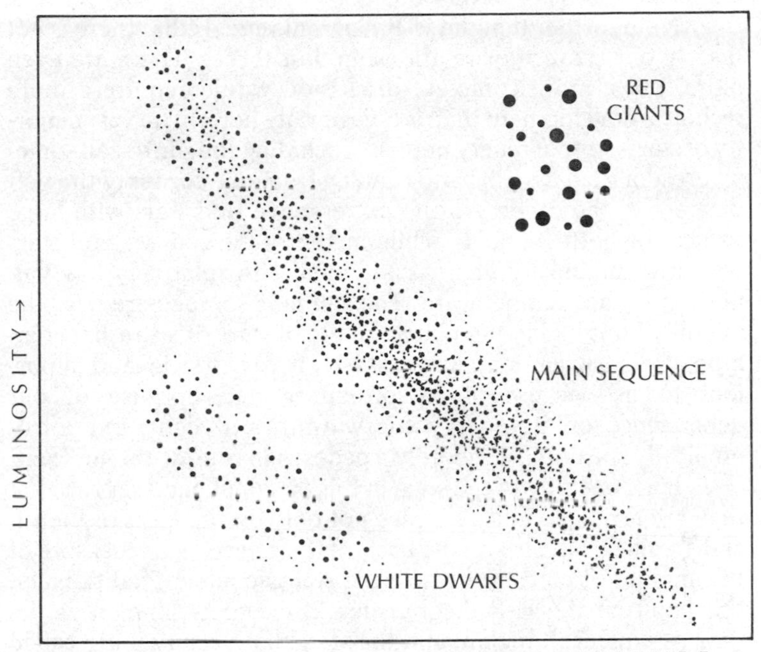

FIGURE 7-3. A REPRESENTATIVE HERTZSPRUNG-RUSSELL DIAGRAM. The points on this H-R diagram represent stars positioned according to their luminosity and temperature values. The vertical axis marks luminosity values increasing from bottom to top, and the horizontal axis marks surface temperature values increasing from right to left (the hotter the stellar surface, the farther to the left it will be placed on the diagram). This simplified diagram shows three of the most common families into which observed stars typically fall when charted in this fashion.

7-3; several of its features deserve comment. Note, for example, that the distribution of points—each representing a star according to its L and T values—is not random, but displays a distinct pattern. This indicates that some combinations of L and T are common while others are only rarely if ever found. This implies that the values of the surface temperature and the luminosity of a star are related; there is something about the behavior of a star that leads it to have a particular combination of L and T values. We will explore that more fully in the next chapter when we discuss stellar behavior, but for now let's just note that stars display a patterned relationship between their luminosity and temperature values.

Note further that the H-R diagram reveals that there is not merely one pattern or relationship, but three. (There are even more, but a more complete discussion would require a more technical development than is appropriate here.) The vast majority of stars—about ninety percent (including the sun)—fall somewhere along a diagonal band running from one corner of the H-R diagram to the other. At one extreme we find stars with large values for both L and T, while at the other end we find stars with low luminosity values combined with relatively low values for surface temperature. The sun falls somewhere near the middle of this highly populated family of stars; it's a rather common, ordinary, garden-variety star. It is very special and important to us because of its proximity and because of our dependence on its energy for warmth and light, but astronomically speaking, it's merely a pedestrian, run-of-the-mill star.

Further study of the stars in this diagonal band reveals that they are arranged in a sequence not only on the basis of their L and T values, but also on the basis of their mass—the measure of the amount of matter of which the star is made.[6] Traditionally, this band on the H-R diagram has come to be known as the "main-sequence" region, and the stars that occupy it are called "main-sequence" stars.[7] The mass values of main-sequence stars lie within the range of approximately one-twentieth to fifty times the mass of the sun, which has a mass of more than 300,000 times the mass of planet earth. The sun, a medium-size main-sequence star, has a diameter of nearly a million miles. As determined with the aid of the Stefan-Boltzmann law, the size of other main-sequence stars ranges from about one-tenth to ten times the size of the sun.

While approximately ninety percent of all stars are main-sequence stars, there are significant numbers of stars in other well-defined and separated regions of the H-R diagram. In the lower left region, for instance, we find a number of stars that are relatively low in luminosity but that have quite high values for surface temperature. Although the mass values for these stars

---

6. The term *mass* has a very specific meaning, distinctly different from both the volume and the weight of a body. "Volume" deals only with the size of the region occupied by the body; "weight" is the gravitational force that this body experiences in the presence of other bodies of mass. "Mass" is a measure of the quantity of matter a body possesses irrespective of the volume it occupies or the gravitational context in which it is found.

7. The main-sequence "region" is strictly a region on the H-R diagram, not a physical region in space.

are similar to the mass value of a typical main-sequence star like the sun, they differ markedly in another property—size. Applying the Stefan-Boltzmann law once again, we find that these stars have a diameter of about 10,000 miles, only one percent of the diameter of the sun. These white-hot stars appear to be something like a sun compressed into the volume of the earth, the volume of which is a million times smaller than the sun's. Their combination of thermal and dimensional properties suggests the title, or family name, by which these stars are known: white dwarfs—"white" because of the color of their hot surfaces, and "dwarf" because of their diminuitive dimensions.

In the other corner of the H-R diagram, the upper right corner, we encounter stars like Betelgeuse and Antares. Light from these stars has a somewhat reddish cast as a result of their comparatively low surface temperature. But in spite of their cooler surfaces, these red stars emit light energy at a staggering rate. Betelgeuse, for example, is a hundred thousand times more luminous than the sun. How can this be? How can a star considerably cooler than the sun emit light energy at a vastly greater rate? By now the answer should be clear: Betelgeuse and stars similar to it are enormous, with huge radiating surfaces. These "red giant" stars, as they have come to be known, are hundreds of times larger than main-sequence stars. Both direct measurement and the Stefan-Boltzmann law indicate that Betelgeuse has a diameter of roughly 700 million miles. This is so large that if Betelgeuse were to replace the sun, the planets Mercury, Venus, Earth, and Mars in their natural orbits would lie within its surface. We would find ourselves surrounded by a red-hot inferno and would soon vanish into stellar vapors—a scenario not so different from some poetic representations of the consummation of the created cosmos.

When I said earlier that the H-R diagram reveals that a star is not just a star, I was pointing to the fact that a star is always some particular kind of star—not merely a standard-issue celestial luminary but always a member of a particular family of luminaries. A star is not just a star; it is a main-sequence star or a white dwarf or a red giant or some other specific type of star. All stars of a certain type share a number of family characteristics, and the set of properties that characterizes one family is vastly different from the combination of features that identifies another. A star is not just a star; it is a star with a family name—a name that serves as a label to represent its unique combination of stellar properties.

As is so often the case in the natural sciences, one discovery leads to a whole new set of questions—questions of a character and specificity that could not be anticipated, questions that could flow only from the spring of empirical investigation. Observations that provide answers to one set of questions usually generate a list of new and more specific questions for research. Fruitful questions tend to multiply, giving birth to successive generations of inquiry. Scientific investigation will never provide all of the answers, but it does promote the formulation of more precisely stated questions within its domain.

What questions are suggested by the H-R diagram? Questions pertaining to those stellar families. Why, we should ask, are there several different kinds of stars? What causes a star to be a member of one family and not another? Are these families related in some way? If so, in what way? Why is the sun a main-sequence star and not a red giant or a white dwarf? How are main-sequence stars related to red giants and white dwarfs? We will take up these and other questions in Chapter Eight.

CHAPTER EIGHT

The Scientific Investigation of Stellar Behavior and History

WHY DO STARS SHINE?

We have all had the experience of standing outdoors on a clear night and looking up to see the stars twinkling like "diamonds in the sky." But how often have we paused to ask why—why do stars shine?

Stars are visible to us because they are luminous, and, as we have already noted, they are luminous because they are very hot. The question, then, is how they get hot and what source supplies the incredible amounts of energy necessary to sustain their high temperatures over extended periods of time.

Ordinary Burning?

Could ordinary burning—a chemical process—be the answer? That possibility quite naturally comes to mind when we observe the rising and setting sun; it does indeed look like a ball of fire. To test whether or not the surface of a star could literally be on fire is relatively straightforward. We begin by asking what materials and conditions would be necessary. All fires need two ingredients: fuel and oxygen. Does a star possess the necessary fuel? It certainly does. A star is composed mostly of hydrogen, which makes an excellent fuel. Is there also a supply of oxygen available? Definitely not. There is not enough oxygen present in or near a star to support chemical combustion for more than a brief instant—no more than a flicker would result. Thus we conclude that in spite of its appearance to us, the sun is not a ball of fire. Its surface, or the surface of any other star, may be as hot as fire (even hotter), but there is no way that it could be literally burning. Some other process must be occurring to produce a star's high surface temperature and luminosity.

Gravitational Collapse?

About a century ago another answer to our question was proposed by Kelvin and Helmholtz. Could the heat and luminosity of the sun and other stars be generated by the process of gravitational collapse and contraction? Under certain specifiable conditions, a massive cloud of gas, such as those found within gigantic interstellar nebulae, will collapse under its own weight. The total gravitational attraction of all the matter in the cloud will overcome all opposing forces, and all parts of the cloud will begin to compress together. Any time a gas is compressed into a smaller volume, it tends to heat up. Computations performed by Kelvin and Helmholtz showed that the gravitational collapse of a cloud of gas having a mass comparable to that of a star would inevitably lead to the formation of a hot ball of gas the dimensions, surface temperature, and luminosity of which would be similar in value to those for a typical star. Gravitational collapse and contraction, in other words, produces a star-like object.

But does gravitational collapse provide an adequate energy supply for a star? To be adequate, the energy supply must satisfy two requirements: it must provide energy at the proper rate (to match a star's luminosity) and for the appropriate length of time. The matter of rate appeared to present no difficulty, but the longevity requirements introduced a vexing puzzle.

How long must a star's luminosity last? The star that provides us with the clearest and most direct physical evidence for calculating its age is the sun. How old is the sun? How long has it had a luminosity comparable to its present value? To answer this question astronomers must enlist the aid of their colleagues in geology. Already at the beginning of the twentieth century, geologists, using a variety of lines of evidence, concluded that the earth and other members of the solar system had been in existence for several billions of years. Current computations, based on evidence drawn from studies of the earth, the moon, and numerous meteorites, suggest that the solar system is about

4.6 billion years old.[1] During that time the earth's surface has experienced a number of significant changes, but the sun has remained relatively stable.

Could gravitational collapse provide enough energy to maintain the sun's luminosity for several billions of years? A relatively simple computation reveals that it definitely could not. The gravitational energy supply is insufficient by a factor of more than a hundred; it could not have kept the sun shining for even one percent of its current age. Gravitational collapse, therefore, does not provide a satisfactory answer to the question of why stars shine. It may provide a plausible explanation of how stars are formed but not for how they keep shining.

Thermonuclear Fusion?

Until a half century ago, no processes other than combustion and collapse were known that might account for stellar energy production. But since that time we have learned about another likely candidate: thermonuclear fusion—the process that has been applied destructively in the hydrogen bomb.[2]

Thermonuclear fusion is the process in which two or more of the smaller atomic nuclei (such as the nuclei of hydrogen or helium) are joined together, or fused, to form a larger nucleus, releasing prodigious amounts of energy as a by-product. Since the nuclear forces that bind atomic nuclei together act only over extremely short distances, nuclei will not fuse unless they are first brought very close to one another. That requires some effort, because the electrically charged nuclei ordinarily tend to repel one another, reducing the probability of fusion. (Actually, that is a very fortunate situation for us. The majority of atoms in our bodies are hydrogen, and surely we would not take kindly to the possibility of our bodies being on the brink of functioning like a hydrogen bomb.) We can bring two nuclei close to one another in the laboratory with a device called a particle acceler-

---

1. For an excellent discussion of the age and formation of the solar system, see William K. Hartmann, *Moons and Planets,* 2d ed. (Belmont, Cal.: Wadsworth, 1983), pp. 119-53.

2. This device ought not to be confused with its little brother, the fission bomb, sometimes called the "atomic" bomb. Nuclear fission is the process in which larger nuclei are broken down into smaller ones. The energy released by this process can be applied destructively, as in weapons, or constructively, as in the modern nuclear power plant.

ator, in which nuclei are accelerated to high speeds and smashed into one another. High speeds (a significant fraction of the speed of light) are necessary to overcome the natural electrostatic repulsion of the like-charged nuclei. In a low-speed collision, the nuclei are deflected away from one another before they come close enough for the short-ranged attractive nuclear forces to promote fusion. In a high-speed collision, however, some nuclei may get close enough to fuse before the repulsive electrical force has a chance to prevent it.

But how might such collisions be arranged in the interior of a star? Quite obviously, there is nothing so artificial as a particle accelerator there. But there is another way to accelerate particles to the sort of high speeds necessary for fusion—by heating them. The higher the temperature of a gas, the higher the speed at which its particles move. If hydrogen gas, for example, were heated to a temperature of 10 million °K, the electrons would be stripped away from the atoms, and the nuclei would travel at speeds sufficient to encourage fusion. When the high speeds necessary for nuclear fusion are generated thermally—by heating—the process is called "thermonuclear fusion." If, then, the hydrogen in the interior of a star could be heated to temperatures near 10 million °K, thermonuclear fusion would occur naturally, and that would generate the energy necessary to maintain the star's luminosity over long periods of time.

But does it happen? Does thermonuclear fusion occur in the interior of stars? It will if all of the necessary conditions are met. One rather obvious condition is that stars must possess an abundance of the proper kind of nuclei. And they certainly do; more than ninety percent of the atoms in a typical star are hydrogen—the perfect thermonuclear fuel. The second condition is that there must be a means of compressing the hydrogen to sufficiently high values of both density and temperature. High density is required so that nuclear collisions will occur frequently; high temperature assures that collisions will bring nuclei sufficiently close for fusion. According to contemporary astrophysics, gravitational collapse will produce both these effects.

The discovery that the processes of gravitational compression and thermonuclear fusion play principal roles in the life history of stars represents one of the major accomplishments of twentieth-century astronomy. With that in mind, let's look at our present understanding of stellar evolution—the sequence of physical processes that brings a star from its "birth" to its "death."

STELLAR EVOLUTION

Stars Do Change

To the casual observer, stars are magnificent symbols (should we say shining examples?) of constancy. Their locations on the celestial sphere appear rigidly fixed. Night after night they assume the same positions relative to one another. Year after year they are seen at the same place in the sky at any specified moment. Such changelessness and constancy have inspired peasants, poets, priests, and philosophers for centuries.

To the astronomer, however, a star is an object that must necessarily exhibit changes in its physical properties. One need not even know the details of those changes to understand why some change is necessary. A relatively simple argument based on energy conservation will suffice to illustrate the point.

The very fact that stars are luminous requires that stars be undergoing change. Stars emit energy in the form of light—that's why we're able to see them. But that light energy must be the product of some physical process by which some other form of energy is being converted into light. That in itself implies a process of change. Some source of energy is being consumed in order that a star may emit light.

Not only can we conclude that stars must be changing, but we can also be certain that the lifetime of a star is finite. Stars cannot remain luminous forever, because they necessarily have limited amounts of fuel, in whatever form it happens to be. Sooner or later that energy source will become exhausted, and the star must "die"—it must cease being luminous. Likewise, the stars that we now see must have been "born" at some time in the finite past. Stars cannot be infinitely old.

Processes of change within a finite lifetime are a necessary part of a star's behavior. The complete story of a star's temporal development begins with stellar birth, follows its development into maturity, and ends in the exhaustion of its energy resources. This full life-history of a star is what astronomers call "stellar evolution." Stars do change, and stellar evolution is the sequence of changes that take place in them as a result of the physical processes that occur both on their surface and in their interior. We might call it the "physical life history" of a star—the story of its birth, development, and death. It is to the investigation of this history that we devote the remainder of this chapter.[3]

How Stellar Evolution Is Studied

But how do we go about studying the life history of stars? It's not as if we could simply observe and measure the properties of a star at periodic intervals and record the changes. Such an approach would work well for studying the life cycles of frogs and fruitflies, but it rarely works for stars. While some stars are occasionally observed to undergo measurable changes, most events in the life history of stars occur so slowly that the purely observational approach just won't work. The time scale for the evolution of a typical star involves not merely the years or tens of years available to an astronomer but rather periods measured in millions or billions of years.

Because the lifetime of stars is so long relative to the "threescore and ten" allotted to astronomers, a theoretical and computational method of investigation must be used in the study of stellar evolution. The modern high-speed digital computer has been employed to simulate the behavior of stars. The basis for the computational program is the assumption that the material of which stars are made behaves according to the patterns described by all of the known physical laws. We assume, for example, that all of the laws for motion, gravity, radiation, thermodynamics, and nuclear reactions apply to stellar material in the same way as they apply to matter investigated in earthbound laboratories.

Having made the standard assumptions of natural science, astronomers have during the past few decades been able to compute the consequences of these assumptions for stellar be-

---

3. Our discussion of stellar evolution is necessarily brief. Almost any textbook in general astronomy will provide a more detailed account. Excellent examples of such texts are Michael Zeilik's *Astronomy: The Evolving Universe,* 3d ed. (New York: Harper & Row, 1982), George O. Abell's *Exploration of the Universe,* 4th ed. (New York: CBS College Publishing, 1982), William J. Kaufmann III's *Universe* (New York: W. H. Freeman, 1985), and Michael A. Seeds's *Foundations of Astronomy* (Belmont, Cal.: Wadsworth, 1984). For discussions focussed on stellar evolution alone, see William J. Kaufmann III's *Stars and Nebulas* (San Francisco: W. H. Freeman, 1978) and A. J. Meadows's *Stellar Evolution,* 2d ed. (New York: Pergamon Press, 1978).

havior. They have been able to assemble a remarkably coherent picture of stellar history. Of course the test of any computational model lies in actual observation; as we noted earlier, only if the results of computation and observation agree do we gain confidence in the theoretical models for stellar behavior and history.

STELLAR GENESIS

In the spiral arms of the Milky Way galaxy (and other galaxies as well), we find numerous large clouds of gas (mostly hydrogen) and dust. We call these tenuous giants "interstellar nebulae." Luminous interstellar nebulae are some of the most beautiful subjects for telescopic photography. (See, for example, the photograph of the Great Nebula in Orion on the cover of this book.) They are usually accompanied by a number of very luminous hot stars that produce a lot of ultraviolet radiation, which is absorbed by the hydrogen atoms in the nebula and reradiated as visible light. The hydrogen emission-line spectrum is dominated by a strong line in the red portion of the visible spectrum, which means that these nebulae ordinarily appear to glow with a deep red light. Mixed with the red, we sometimes also find some blue light scattered by the dust particles in the nebula. This blue is produced by the same process that makes our own sky appear blue from scattered sunlight.

The typical nebula is tens or hundreds of light-years across and contains as much material as many thousands of suns. Within the cloud the mass is neither uniformly distributed nor static. In response to various forces, the gas moves about, sometimes forming regions in which the material is concentrated into well-defined "globules" of gas and dust.

Under the right conditions, these globules may become stellar embryos or "protostars." When a sufficient amount of gas at a low temperature gathers into a small enough volume, it necessarily becomes subject to the processes of gravitational collapse and contraction. The collapsing protostar is not yet a star, but it is on the way to becoming one. The beginning of the gravitational collapse process is like cresting the first hill on a roller coaster ride. Once started, there is no turning back until the whole ride is over.

Initially, the collapse proceeds quite rapidly. In a period of a few decades or centuries (the specific time depends on the initial size, mass, and temperature of the globule), the collapse reduces the globule to a hot ball of gas about a hundred million

M 16, the Eagle Nebula, seen in the direction of the constellation Serpens. The dark structures visible in this photograph are relatively dense "globules" of gas and dust that block our view of the luminous material behind them. These structures are candidates for the formation of stars by the gravitational collapse process.

miles in diameter—about a hundred times larger than the sun. At this stage the gas pressure within the protostar has become sufficiently high to support the weight of the outer regions, much like the earth's atmospheric pressure prevents the further gravitational collapse of of the air we breathe. The rapid collapse phase of the protostar is thereby brought to a halt as further collapse is opposed by gas pressure. The product of rapid collapse is a ball of gas with a surface temperature of a few thousand degrees K and a central temperature about ten times higher.

Such a hot ball of gas is luminous. The source of the energy that is being emitted as light is the gravitational potential energy that was lost during collapse. After the rapid collapse, this transformation of gravitational energy continues, but at a diminished pace. The protostar then goes through an extended period of gradual gravitational contraction. For a protostar with a mass similar to that of the sun, this process will last for a few tens of millions of years. During this period of contraction, about half of the diminished gravitational energy is transformed into radiation and the other half is transformed into heat. Most of that heat builds in the stellar core—the central region, which is most compressed by the weight of the surrounding matter.

As a result of the gradual gravitational contraction, the core of the protostar is both compressed and heated. Computations indicate that a core temperature of 10 million °K will be reached as long as the protostar has a mass of at least one-tenth of the mass of the sun. As we have already noted, this is the temperature at which thermonuclear fusion will occur. Thus, after rapid gravitational collapse and gradual gravitational contraction, the process of thermonuclear fusion will inevitably begin. The protostar has no choice in this matter. Thermonuclear fusion is the necessary consequence of the process of gravitational collapse and contraction.

The onset of fusion in the stellar core introduces a significant change in the state of affairs. Before this point, the heat and luminosity of the protostar were entirely dependent on the dwindling supply of gravitational energy. But now the fusion process, which draws on the huge resources of nuclear energy, takes over the responsibility for maintaining the high temperature and luminosity of the protostar. In fact, the heat generated by the fusion process in the core accomplishes two things: it increases the thermal gas pressure to such a degree that further gravitational contraction is halted, and it provides just the right amount of energy to maintain the star's luminosity. The energy produced by thermonuclear fusion slowly works its way from the hot stellar core to the cooler surface, from which it is radiated as starlight.

Once the fusion rate has built up to the proper level so that contraction ceases and a steady luminosity is achieved, we have an extremely stable object. Its size is no longer changing—the gravitational force inward is exactly balanced by thermal gas pressure outward. Its luminosity is being maintained by the fusion process, which consumes a relatively small amount of fuel to produce very large amounts of energy. One way to describe this combination of conditions is to say that an *equilibrium* has been established. The star is, for the time being, stable—steady in size, in temperature, and in luminosity.

But what is this "equilibrium thing"? (Or shall we call it an E.T.?) Can we recognize this product of collapse, contraction, fusion, and equilibrium as truly a star? If so, what kind of a star is it?

To answer this series of questions, it will be extremely helpful to make use of a luminosity-temperature diagram like the

Hertzsprung-Russell diagram I introduced in Chapter Seven. Let's begin by tracing the history of a representative protostar from globule to "E.T." Initially, the globule is cold and hardly shining, which places it in the lower right-hand portion of the L-T diagram (see figure 8-1). Rapid collapse produces an increase in both temperature and luminosity, causing the position of the protostar to move up and toward the left on the diagram. The line traced out by this continuously changing position is called the "evolutionary track" of the protostar.

When rapid collapse is replaced by gradual contraction, the evolutionary track takes a nearly vertical drop. Though the internal temperature of the protostar is increasing significantly, the growing opacity of the stellar material shields the surface from this heat, and the surface temperature remains relatively constant during contraction. If its temperature remains constant, a surface that is diminishing in area will decrease in total luminosity as well. Thus, the evolutionary track drops downward on the diagram.

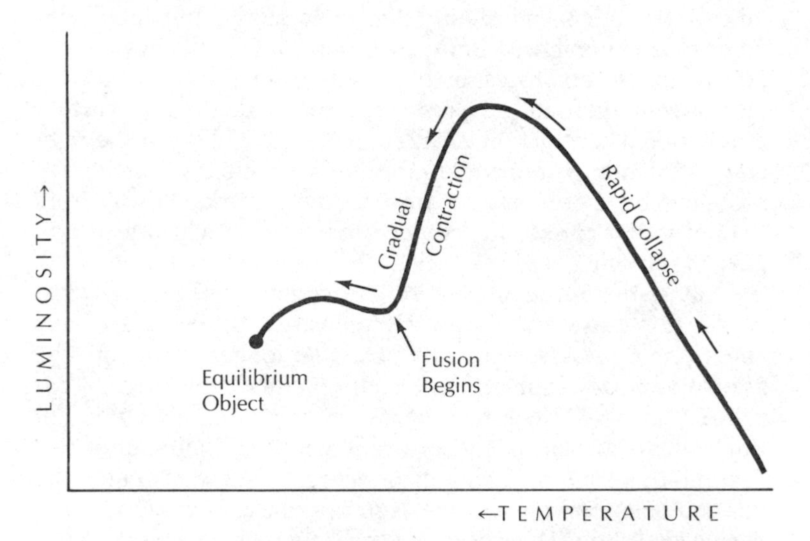

FIGURE 8-1. A TYPICAL EVOLUTIONARY TRACK FOR A PROTOSTAR. During a relatively brief period of rapid gravitational collapse, a globule of gas and dust will build to a peak of luminosity. During the long period of gradual contraction that follows, the protostar's luminosity diminishes as the gravitational potential energy is expended. Eventually, when the necessary conditions are established, thermonuclear fusion begins in the protostar's core and a state of equilibrium is established: a star is born.

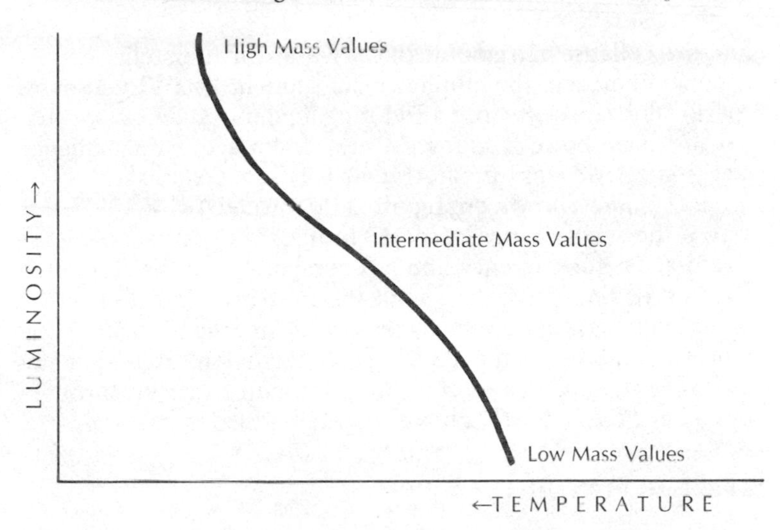

FIGURE 8-2. THE ZERO-AGE MAIN-SEQUENCE LINE. This line is formed by plotting the locations of "equilibrium things" having mass values from one-twentieth to fifty times the mass of the sun. It is called the Zero-Age Main-Sequence line because with the aid of this diagram we are able to recognize the "equilibrium things" as newly born ("zero-age") main sequence stars.

When fusion begins, there appears another kink in the evolutionary track as the protostar adjusts to its new source of energy transformation. During this adjustment period, as the contraction comes to a halt and equilibrium becomes established, the surface temperature increases, although the luminosity undergoes only minor changes. Once equilibrium is established, the position of the E.T. is quite stable.

The location of an E.T. on the luminosity-temperature diagram depends principally on the value of its mass. It is possible to compute the locations for objects having the full range of stellar mass values. Figure 8-2 illustrates the results of such computations.

Two features of figure 8-2 should strike you as quite remarkable. First, the distribution of equilibrium objects on the L-T diagram is clearly the same as that exhibited by main-sequence stars. Second, the particular way in which the mass value fixes the position of an equilibrium object along this line is precisely the same as the way mass fixes the location of stars along the main sequence. We can conclude, therefore, that our "equilibrium thing" is a main-sequence star. The product of the gravitational collapse of a globule of gas within an interstellar nebula is none other than the familiar main-sequence star. Even more specifically, since we are considering a main-sequence star that has just been born, we may call it a "zero-age main-sequence star"—a ZAMS star, for short. Our E.T. is a ZAMS star.

The line shown on figure 8-2, formed by ZAMS stars across the whole range of possible mass values, is called the "zero-age main-sequence line." A comparison of the computed position of the ZAMS line with the measured location of the main-sequence region reveals that the ZAMS line lies along the lower boundary of the main-sequence region. Actual main-sequence stars, including the sun, are found scattered throughout a band immediately above the ZAMS line.

STELLAR MATURITY

Life on the Main Sequence

Once equilibrium has been established, and a ZAMS star has been formed, the pace of stellar development slows down considerably. It does not come to a standstill, however. Even stable main-sequence stars must change, as our earlier argument from energy conservation demonstrated. Specifically, within the core of a main-sequence star, the thermonuclear fusion process is gradually transforming hydrogen into helium.

During the main-sequence phase, which is the longest segment of a star's history, its surface temperature remains relatively constant, but its size and luminosity increase very slowly. On the luminosity-temperature diagram, the star proceeds at a snail's pace upward through the main-sequence region.

The end of the main-sequence phase is brought about by changes in the core region of the star. The helium that is produced by the fusion process collects in the core, where it was formed. When approximately ten percent of a star's hydrogen has been transformed into helium, the accumulated helium in the core triggers a relatively abrupt change in the rate of energy production, and the star leaves the main-sequence region—it ceases being a main-sequence star and becomes a star of a different type, which we will discuss shortly.

The details of a star's physical and structural changes are not particularly important to this discussion, but the timing of events in the life-history of a star is of considerable interest.

Fortunately, the computational models for stellar evolution allow us not only to construct evolutionary tracks but also to determine the time scale for evolutionary processes. The time period most characteristic of stellar history is the amount of time a star spends in its main-sequence phase. Astronomers call this the star's "main-sequence lifetime."

The length of a star's main-sequence lifetime depends very critically on the amount of its mass. For a star with a mass comparable to that of the sun, the main-sequence lifetime will last approximately ten billion years. In general, less massive stars have considerably *longer* lifetimes, and more massive stars have *shorter* lives. Although very massive stars have greater amounts of fuel to consume, they expend it at such a high rate that their lifetimes are relatively brief in stellar terms. (In a peculiar sense, stars are like cars: the heavier they are, the poorer their gas mileage.) Table 8-1 shows the approximate main-sequence lifetime values for stars of different mass values.

Where is the sun in its life history? According to empirical evidence gathered by both astronomers and geologists, the sun and the solar system are approximately 4.6 billion years old.

MASS VALUE	MAIN SEQUENCE LIFETIME VALUE
30	2 million years
10	30 million years
3	600 million years
1 (the sun)	10 billion years
⅓	200 billion years
⅒	3 trillion years

TABLE 8-1. The life expectancies for main-sequence stars of different masses. Mass values are expressed in multiples of the sun's mass (which is approximately 332,000 times that of the earth).

Stellar evolution computations indicate that the main-sequence lifetime of the sun is ten billion years. Thus, we conclude that the sun is about halfway between its birth and its death. Or, to put it another way, the sun is a middle-aged star.

Stellar Senior Citizens

As the main-sequence star grows older, helium accumulates in its core, eventually bringing the main-sequence phase to an end. When the core has become nearly pure helium, hydrogen fusion can no longer occur there; it is forced to take place in a spherical shell surrounding the core. When fusion stops in the core, the core will resume the process of gravitational contraction. As a consequence of contraction, the core heats up. This in turn heats up the shell in which hydrogen fusion is taking place, causing the fusion to proceed at a much accelerated pace.

This sequence of events has the net effect of bringing about a significant increase in the star's rate of energy production. When the energy being produced in the core by consuming nuclear fuel at the new higher rate eventually works its way to the surface of the star, the star manifests a dramatic increase in luminosity. Accompanying the luminosity increase are two other major changes in stellar properties: the size of the star is increased about a hundred-fold, and the greatly expanded surface cools to roughly half of its original temperature.

The product of all of these changes is a star that is highly luminous, hundreds of millions of miles in diameter, and relatively red because of its reduced surface temperature. Its identity should come as no surprise—it is none other than the red giant star. Originally discovered through observation, the red giant has now been "discovered" and accounted for by means of mathematical computation as well. As a consequence of ordinary physical processes, main-sequence stars undergo structural and physical changes that transform them into red giants, the senior citizens of the stellar population.

STELLAR DEATH

The story of stellar history through the red giant phase has been a single story; that is, the history of all stars is essentially the same, irrespective of the amount of mass of which they are made. The only effect of mass that we needed to consider was its influence on the time scale: the greater the mass, the more quickly a star proceeds from birth to maturity and senior citizenship.

Beyond the red giant phase, however, the subsequent history of stars is very dependent on their mass value. Stars with large mass values proceed through vastly different states and events than stars with small mass values. In each case, a star is faced with the exhaustion of its energy reserves and the relentless force of gravity, which seeks to crush it into ever diminishing volumes. But differences arise as a result of the way the mass value determines the internal temperatures achieved in a dying star. Greater mass causes stronger gravitational compression forces. The degree of gravitational compression governs the highest temperature that can be generated in the contracting core.

The temperature value in a stellar core is important because it determines which fusion reactions are permitted to take place. As we discussed earlier, high temperature is necessary in order to promote collisions between nuclei at sufficiently high speeds to overcome the electrostatic repulsion between like-charged nuclei. Larger nuclei, which are more highly charged, require higher temperatures than smaller ones. You will recall that hydrogen nuclei will fuse to form helium nuclei at a temperature of 10 million °K, but it takes a temperature of 100 million °K before helium nuclei will fuse to form carbon nuclei. Larger nuclei require still higher temperatures, even billions of degrees K. Since stellar mass determines the temperature that can be achieved in the contracting core, it also determines how far the sequence of thermonuclear fusion reactions will proceed.

Stars with less than half the mass of the sun never achieve a central temperature sufficient for the fusion of helium into heavier elements. According to computation, after such a low-mass star completes its red giant phase, it will slowly cool off and shrink into a planet-sized remnant of the original star. Old low-mass stars don't really die, they just fade away into the oblivion of invisibility.

In more massive stars, like the sun, the fusion of helium into carbon and some heavier elements will occur. But there will inevitably come a time when the conditions (either fuel supply or temperature) for fusion will no longer be satisfied. In many cases, the star will be divided into two parts. The gravitationally collapsed core will be left behind while the outer envelope of the star is ejected into space like a cosmic smoke ring, forming what astronomers call a "planetary nebula."[4] The remnant left behind is like the fading remnant of the less massive stars—a small (about the size of earth) but dense ball of material in which fusion no longer occurs. It glows with the energy left from previous events, supplemented with some gravitational energy as it continues to contract slightly. The combination of its properties and behavior allows a positive identification of this object: it is a white dwarf star.

Those red giants with a mass greater than two or three times the mass of the sun can support a still longer chain of nuclear fusion reactions. The products of one reaction serve as the fuel for the fusion of even heavier elements. The fusion of progressively heavier elements, however, produces progressively less energy than did the fusion of hydrogen into helium. And the pace of fusion must increase markedly in a frantic rush to supply the energy for a star's luminosity. Helium fuses into carbon and oxygen. From these are formed neon and magnesium. Further fusion yields silicon, sulfur, and iron. But iron is unique. Iron has a quenching effect on the nuclear activity. Both the fusion of iron into heavier nuclei and the fission of iron into smaller nuclei result not in energy production, but rather in energy absorption. An iron core in a star behaves like a bucket of water on a bonfire.

The quenching action of the iron core of a massive star triggers a dramatic burst of activity. In an attempt to replace the energy absorbed by the iron nuclei, the core gravitationally collapses into a tiny ball. The outer regions rush toward the vacated central area and are suddenly subjected to violent forces generated by both thermonuclear fusion and gravitational energy conversion. A significant fraction of the star is explosively expelled in this event, called a "supernova." In a matter of hours the luminosity of an exploding star can increase by a factor of billions, rivaling the luminosity of an entire galaxy for a short time. The gravitationally collapsed core is left behind—a stellar corpse crushed into a rapidly rotating ball of neutrons called a "neutron star," or a "pulsar," a star that emits bursts of energy in periodic pulses.

There is one feature common to all of these scenarios for

---

4. The term "planetary nebula" is somewhat misleading. In early telescopic observations, such objects looked vaguely like distant planets—both appeared as round blobs. From that accidental similarity, the name was established. These nebulae have no physical relationship to planets, however.

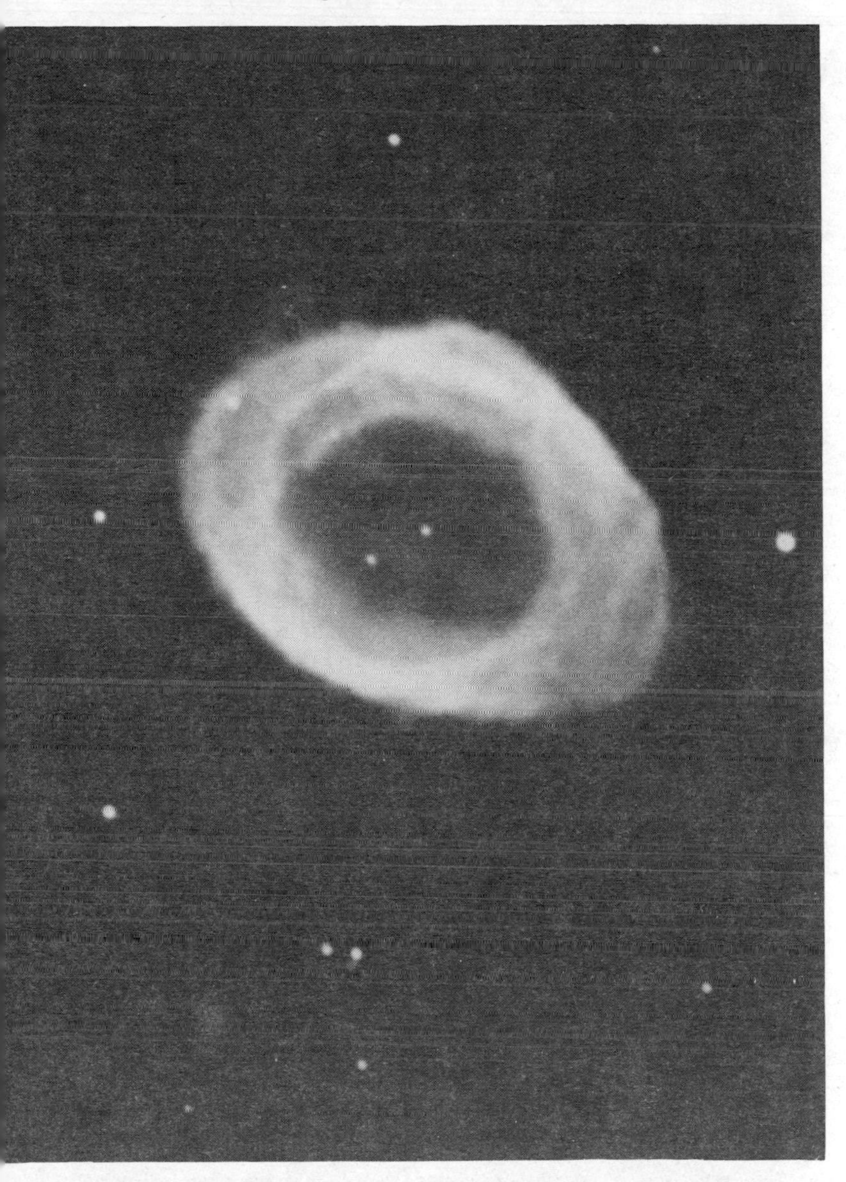

The Ring Nebula, M 57, a planetary nebula in the constellation Lyra. This spherical shell of fluorescent gas is expanding outward from a small hot star. According to contemporary astrophysics, such a structure is given off by an aging red giant star as thermonuclear fusion reactions approach their termination. The remaining central star becomes a white dwarf.

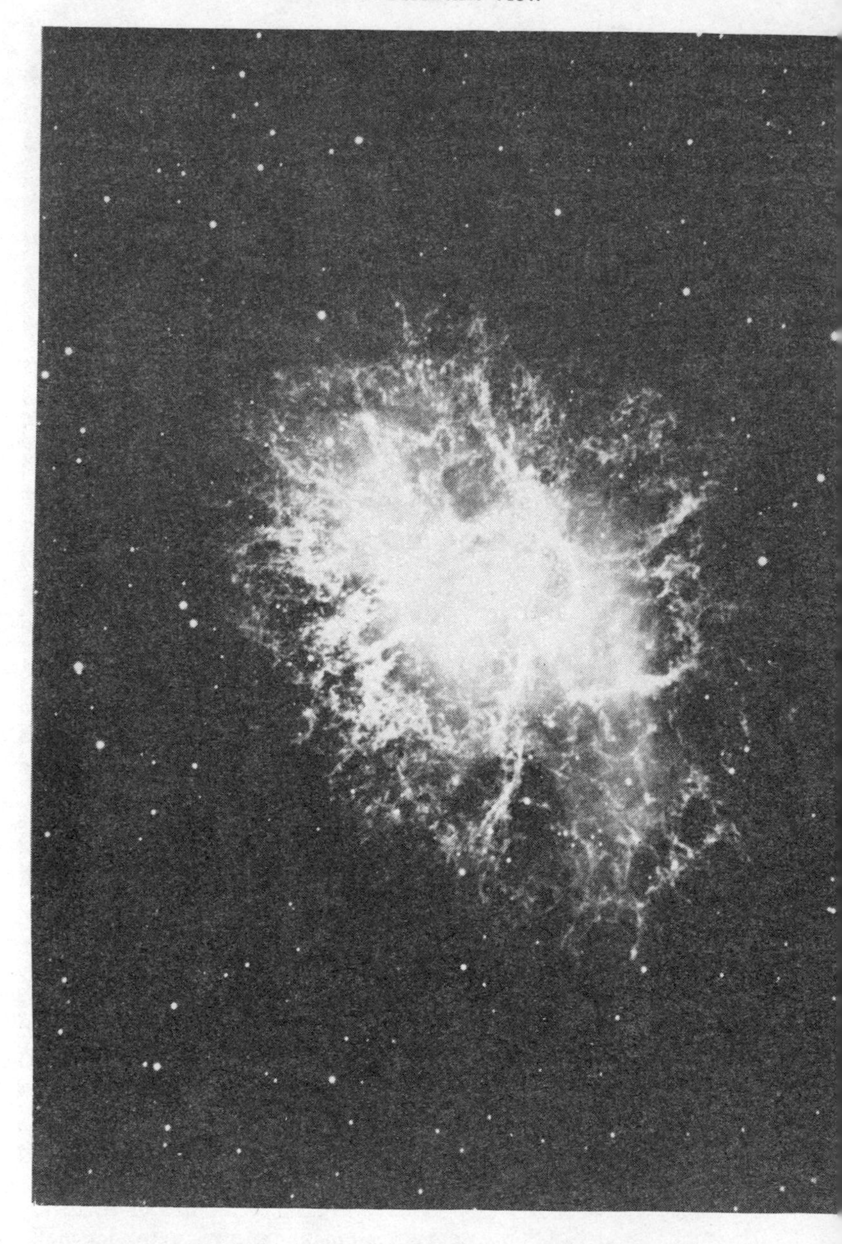

M 1, the Crab Nebula, a remnant of the supernova explosion observed in A.D. 1054. Exploding stars such as this one contribute vast amounts of heavy elements to the interstellar medium—the elements necessary for the formation of planets, pebbles, penguins, and people.

stellar death: in each case a battle is lost. During its entire lifetime, a star is engaged in a battle with gravity. Gravity, which formed the star in the first place, is unrelenting in its efforts to crush the star into nothingness. But as long as thermonuclear fusion can keep up the temperature in the stellar interior, as it does in a main-sequence star, the inward force of gravity will be balanced by the outward force of thermal pressure.

In comparison, consider a typical planet like earth. Here also, gravity provides a centrally directed force on every speck of the planet, on every single atom. So why doesn't the earth simply collapse under its own weight? Because there is an opposing force, a force that arises when atoms are brought very close to one another. If you try to squeeze them into the same volume, interatomic forces arise to resist that action. For this reason none of us need fear that we are in danger of falling into the center of the earth. Gravity will lose that particular battle because of the effective opposition provided by interatomic forces.

In a white dwarf star, however, gravity has already won the battles against both thermal pressure and interatomic forces. Because thermonuclear fusion has ceased, thermal pressure is insufficient to prevent further gravitational collapse. Similarly, the interatomic force is incapable of balancing the tremendous force of gravity within a white dwarf—a star in which a mass comparable to that of the sun is compressed into a ball the size of planet earth. Atoms are literally crushed. In their place we find a mixture of nuclei and electrons that are no longer structured into distinct atoms. The density of this material is phenomenally high—millions of times greater than the density of water. Just one teaspoonful of white dwarf material would weigh ten tons on the surface of the earth.

At the surface of a white dwarf star the forces of gravity are extremely strong. If we could stand on its surface, we would not only find it a bit hot there, but would also discover, if only briefly, that we weighed several thousands of tons each; we would soon enough be reduced to little more than a thin smudge on that surface.

If neither thermal pressure nor interatomic forces are sufficient, what force does prevent the further collapse of a white dwarf? It turns out to be a force associated with electrons (called the "electron degeneracy force"); they have a natural resistance to being confined in very small regions of space, as in the highly compressed white dwarf material. From the observed properties of white dwarf stars and the known behavior of this force, astronomers can be certain that the electron degeneracy force is responsible for stabilizing white dwarf stars against further gravitational crushing.

But there is a limit to the resistance that this force can provide. That, in turn, establishes a maximum amount of mass that a white dwarf star can have—it's called the Chandrasekhar limit. Collapsed stellar remnants with a mass exceeding about 1.4 times the mass of the sun cannot be stabilized at the white dwarf phase. Gravity will win yet one more battle in these massive stars.

Such is the case in the collapse of a massive star preceding a supernova explosion. That collapse cannot be halted by the electron degeneracy force; gravity proceeds to crush the stellar core to the incredible density of a hundred million million times the density of water, comparable to the density of the atomic nucleus itself. In this process even the nuclei and electrons are crushed out of existence, and in their place we find only neutrons, electrically neutral particles similar in mass to protons. For this reason the collapsed object is called a "neutron star."

On the surface of a neutron star gravity is so strong that even a ping-pong ball would weigh nearly a million tons, and a comparable volume of neutron star material would weigh a hundred billion billion tons. What force could possibly prevent further collapse under such weights? Only one: the "neutron degeneracy force." The stalemate between gravity and the neutron degeneracy force represents a star's last defense against the crushing force of gravity. No greater force of resistance is yet known.

But what if the mass of the collapsing star is so great that gravity overcomes even the neutron degeneracy force? As far as we know, gravity accomplishes a complete victory in that case. Gravity apparently wins not only one particular battle but the whole war. An amount of mass several times the mass of the sun can be crushed into an arbitrarily small volume. How small, we don't really know. We are at the limits of our computational and modeling ability. The computations that can be done predict that the mass will be so concentrated that there will be a region of space surrounding it, a few miles in diameter, within which gravity will be so powerful that nothing—not even light—can escape from its grasp. The ultimate gravitational victory is the formation of what has come to be called a "black hole"—black because no light can escape it, and a hole because anything you throw into it disappears from sight.[5] White dwarfs, neutron stars, and black holes all represent the gravitationally collapsed remnants of stars—stellar corpses. Only the amount of mass determines which fate a star will meet. In any case, gravity has won a battle and the luminous life of a star will come to an end.

THE MOST SIGNIFICANT DISCOVERY

Perhaps the most significant aspect of the results of stellar evolution computations is the discovery that the major categories of stars that were first revealed observationally—main-sequence stars, red giants, and white dwarfs—can now be recognized as members of a temporal sequence. Initially, all we knew was that stars came in certain distinct varieties. We asked what might be the relationship among these stellar families, but we had no answer. Now, however, we know that they are related chronologically. Globules collapse to form main-sequence stars. Main-sequence stars eventually change into red giant stars. After the red giant phase, gravity begins to win its battles over opposing forces and crushes stars into white dwarfs, neutron stars, or black holes.

Stars have a calculable life-history. Every star that we see fits into a chronological sequence of material behavior. If we wish to inquire about the temporal development of stars, the natural sciences will provide many answers. Such questions lie within the domain of legitimate scientific investigation. The computation of stellar history represents one of the major accomplishments of twentieth century astronomy. It is a fascinating story the evidence for which is sufficiently compelling and coherent that it clearly ought to find a place in any valid worldview.

OBSERVATIONAL EVIDENCE FOR STELLAR EVOLUTION

Most of the discussion about stellar evolution that I've presented in this chapter is the product of mathematical computation. This is necessarily the case because, as I stated at the outset, the

5. For further discussion on these fascinating objects, see William J. Kaufmann III's *Black Holes and Warped Spacetime* (San Francisco: W. H. Freeman, 1979).

lengthy timescale of stellar evolution prohibits the direct observation of the life cycle of any particular star.

But how can we be certain that real stars do in fact behave like the computational model stars? Does observational evidence give support to the processes of stellar evolution that I've presented? My answer is an unequivocal Yes. The technical and popular literature is filled with reliable reports concerning the observational evidence for stellar evolution. In fact, the computational models are continuously being tested by comparison with observations. That's the way natural science goes about its business. Let's look briefly now at just a few examples of the observational support for stellar evolution.

The Congruence of Observed and Computed Stellar Properties

Stellar history is the cumulative product of stellar behavior. But the behavior of a star, or of any other material system, is closely correlated with its physical properties. Thus, one fundamental requirement imposed on stellar evolution computations is that the properties of model stars must match the observed properties of actual stars. And indeed such a congruence does exist. Recall, for example, our discussion of the identity of the products of the gravitational collapse and contraction process. Within a specifiable range of mass values, these "equilibrium things" were computed to be distributed along a line running diagonally across the H-R diagram, with their particular positions fixed by their mass values. Precisely the same properties are observed for actual main-sequence stars. The range of mass values is the same; the luminosity and temperature values are the same; the effect of mass is the same. Therefore, the products of computation can unambiguously be identified with actual objects.

Similar correlations could be pointed out for stars other than main-sequence: red giants with their contracting cores surrounded by shells of fusion activity, white dwarfs stabilized by the electron degeneracy force, neutron stars whose collapse has been halted by the neutron degeneracy force. The existence of neutron stars, in fact, was predicted thirty years before they were actually observed. Even black holes, those bizarre shadows of stars that once blazed in brilliance, provide the best explanation for the observed behavior of certain binary systems. The simple but fundamental point is that time and time again the observed properties of a broad spectrum of stellar types match extremely closely with the computed properties of stars at various stages in their temporal development. Such a correspondence is, I judge, neither an accident nor a fabrication. Rather, I count it as substantive evidence that the computational models for stellar evolution are realistic and worthy of serious consideration. Many details remain to be worked out, and many interesting anomalies are yet to be resolved, but the larger picture is clear: the stars in our night sky represent a population of stars in a wide diversity of evolutionary stages. The massive, the miniscule, the ordinary, the exceptional, the young, and the old are all represented.

Stellar Maternity Wards

If our understanding of the birth of stars is correct, then their birth sites must be in the massive interstellar nebulae. Much evidence supports this proposition. Numerous condensed globules of gas and dust, candidates for gravitational collapse, are found in these nebulae (for an example, see the photo on p. 146). Also embedded within interstellar nebulae are a large number of "hot spots." These heated regions emit most of their radiation in the infrared portion of the spectrum. From Wien's Law we learn that their temperature is typically a few hundred degrees K, which is precisely what one would expect if a newborn star were embedded deep within the nebula. Radiation from the star would be absorbed by the surrounding dust, heating it to a few hundred degrees K. The heated dust would then produce the thermal radiation that is observed by infrared telescopes.

Associated with most interstellar nebulae, and gravitationally bound to them, are a number of stars. Among these, the stars that are especially interesting are those which are sometimes called "blue giants." They are very massive main-sequence stars that exhibit both high luminosity and high surface temperature. Recall, now, that these massive stars have very short main-sequence lifetimes—they don't last long before leaving the main sequence. Therefore, if we see them at all, they must have just recently formed. They must, in other words, be newborn stars.

Stellar infants should still be close to their place of birth; they would not have had sufficient time to wander far. Where are the blue giants (massive stellar infants) found? Only in association with interstellar nebulae that could have recently—during the past few million years, that is—supported their birth. Finding blue giants associated with large nebulae is like finding infants near mothers or nurseries. Quite appropriately, interstellar nebulae are frequently referred to as "stellar maternity wards." They are the sites of stellar birth, in which it is estimated that new stars are being born at the rate of one per year somewhere in the Milky Way.

Star Clusters

Within the Milky Way galaxy, as well as other galaxies, we find a large number of star clusters. These are groups of stars that are spatially clustered together, not mere accidental groupings like the constellations that we happen to observe from the vantage point of earth. Clusters found in the spiral arms of our galaxy contain anywhere from tens to thousands of stars, usually in a rather open, irregular distribution. Among the thousand that have been catalogued, the Pleiades cluster is perhaps the most familiar to us. Because such clusters are found only near the galactic plane, we call them "galactic clusters." Other clusters, found outside of the galactic plane in the "halo" region of the galaxy, typically contain hundreds of thousands of stars in a close, spherically symmetric distribution. Because of their symmetry, they are called "globular clusters." Approximately 150 are known to be associated with the Milky Way galaxy.

Besides being magnificent objects to observe and photograph, stellar clusters are very interesting for their peculiar physical properties. Scientific studies of these clusters typically emphasize either the spatial distribution of their stars or the distribution of their member stars on an H-R diagram. The latter approach is most relevant to this discussion.

Figure 8-3 shows the results of luminosity and temperature measurements for the stars of a representative cluster. (There are some differences between galactic and globular clusters that we can safely ignore here, since they are quite irrelevant to the argument presented.) As expected, the majority of stars lie along the main sequence, but with one very obvious peculiarity: only part of the main sequence is populated. The lower portion of the main sequence is filled with stars up to a particular point, which I have labeled the "turnoff point." Above the turnoff point the main sequence is unpopulated. The massive stars that we might have expected to find there are missing. Instead, we find them strung out from the turnoff point into the red giant portion of the

M 13, the globular star cluster in the constellation Hercules. This cluster, one of more than a hundred associated with the Milky Way galaxy, contains hundreds of thousands of stars and is located about 30,000 light-years from us.

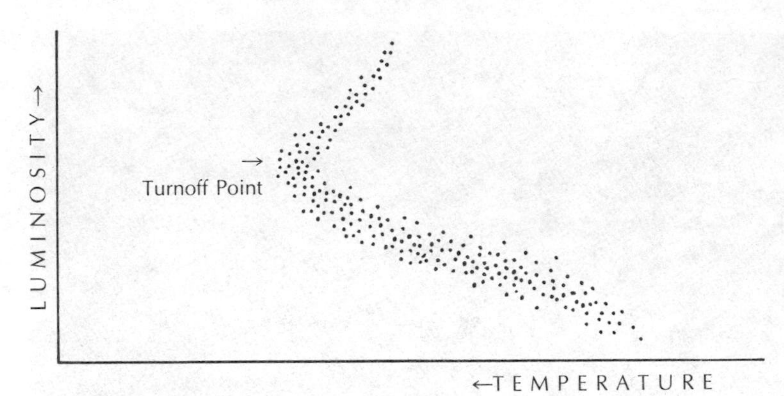

FIGURE 8-3. H-R DIAGRAM OF THE STARS IN A REPRESENTATIVE STELLAR CLUSTER. A comparison of this diagram with the H-R diagram for a random sample of stars (see figure 7-3 on p. 135) reveals a significant difference: on the cluster diagram, that portion of the main sequence that lies above and to the left of the turnoff point is unpopulated—the cluster contains no massive stars in their main-sequence phase.

diagram. Assuming that these differences are not merely accidental, what physical reasons might account for the peculiar properties exhibited by a stellar cluster?

To answer that question, let's first consider a hypothetical cluster formed in a manner most strongly suggested by observational evidence. We must begin by making some reasonable assumptions. First, we assume that the stars of a cluster are not associated by mere accident but rather that they were all born in the same interstellar nebula—the same kind of nebula in which we find evidence of star formation today. Second, let's assume that the majority of stars in a cluster are formed during a relatively short period of time—short, that is, relative to the multibillion year time scale for the evolution of a star like the sun. (There are several mechanisms at work in nebulae that do, as a matter of fact, support our making such an assumption.) Finally, let's assume that among the large number of stars formed from a nebula, no particular mass value was favored; that is, let's assume that initially the newborn cluster of stars contained stars having a normal broad range of mass values.

On the basis of these assumptions, what sort of properties would our hypothetical cluster have immediately after its formation? Drawing on our understanding of star formation, we would expect that at the outset we would find only main-sequence stars in the cluster. In fact, an H-R diagram of an infant cluster should consist of a series of stars distributed along the ZAMS line.[6]

Now, suppose that we left the cluster for a time and allowed the stars to mature in the manner that we have already described. If we came back some time later, what would we expect to find? Suppose, for example, that we came back a billion years later for a second look at our cluster. What changes would have taken place in the meantime? Recall that the rate at which a star changes depends critically on its mass; the greater the mass, the more rapid the changes. Thus, the low-mass stars in our cluster, stars with main-sequence lifetime values that greatly exceed their billion-year age, would still be in their "youth" and would have changed very little. They would still lie along the lower portion of the main sequence. On the other hand, the most massive stars in the original cluster, stars with main-sequence lifetime values considerably less than a billion years, would by now have completed their main-sequence phases and vacated the main-sequence region. Some of these stars might already be in the red giant region, or at least approaching it; others might already have gone to a supernova phase and disappeared from sight.[7]

Our hypothetical cluster, therefore, will yield an H-R diagram that looks essentially identical to those exhibited by actual clusters. The lower main sequence would be fully populated, the upper main sequence would be empty, and some stars would be found between the turnoff point and the red giant region. Figure 8-4 shows the changes that would have occurred during the first billion years of our hypothetical cluster's history. The upper portion of the main sequence appears to be peeled off toward the right as the massive stars vacate the main-sequence region. As the cluster ages, this peeling action continues, and the turnoff point moves progressively further down the main sequence.

The turnoff point on the H-R diagram of a cluster has a very special significance; for one thing, it can provide us with a means

---

6. Strictly speaking, this statement is not correct. Stars with very low masses form much more slowly than massive ones, and so they reach the ZAMS line somewhat later. This effect is quite evident in the H-R diagrams for very young clusters.

7. The neutron star left by a supernova is not always visible. These stars do not radiate uniformly in all directions, and if they don't happen to radiate in the direction of earth, we won't receive any evidence that they are there.

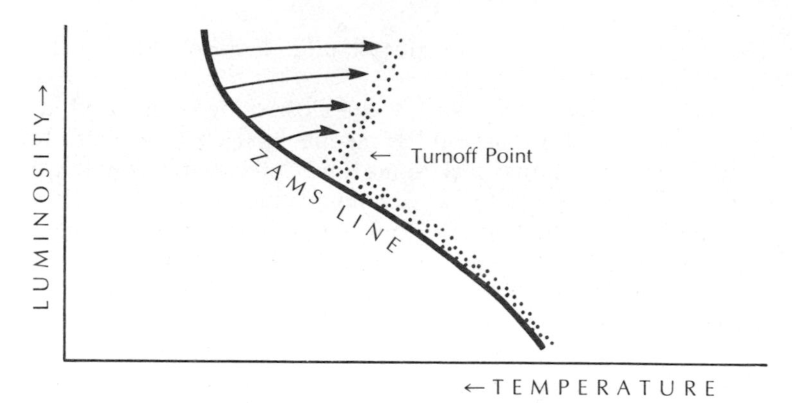

FIGURE 8-4. THE FIRST BILLION YEARS OF CLUSTER EVOLUTION. When the stellar cluster is born, the stars within it will roughly form a ZAMS line when charted on an H-R diagram. By the age of one billion years, however, many of the high-mass stars will have evolved into red giants, leaving behind the longer-lived low-mass stars. As more time passes, the turnoff point will slowly move down the main sequence as more and more stars age and move off toward the red-giant area of the diagram.

of determining the age of the cluster. The massive stars, with main-sequence lifetimes less than the age of the cluster, will already have completed their main-sequence phase and vacated the main-sequence region on the diagram. Less massive stars, with main-sequence lifetimes greater than the age of the cluster, will not yet have completed their main-sequence phase and will therefore remain in the main-sequence region. A star at the turnoff point will be a special borderline case: it will just be completing its main-sequence phase and getting ready to move out of the main-sequence area. Knowing that a star is at that specific stage in its history allows us to determine its age. Its age must be equal to the time that has elapsed from its birth to the end of its main-sequence phase—but that is precisely the definition of its "main-sequence lifetime." We can conclude, therefore, that the age of a star at the turnoff point is equal to its main-sequence lifetime value. All of the stars in a cluster, however, are essentially the same age, since they were born during a relatively brief time period. Thus, the age of the entire cluster is equal to the main-sequence lifetime of a star at the turnoff point on the H-R diagram for that cluster.

Furthermore, the main-sequence lifetime of a star at the turnoff point can be evaluated from its location on the H-R diagram. As we have already seen, the position of a main-sequence star is determined entirely by its mass value. Therefore, the location of the turnoff point identifies the mass value of stars located there. But the mass value also fixes the value of the main-sequence lifetime (see table 8-1). Putting these facts together, it is in principle a relatively straightforward matter to evaluate the main-sequence lifetime of a star at the turnoff point, which will also be, as we have noted, the age of the cluster.

What is the significance of this whole discussion of star clusters? Two points particularly deserve our attention. First, all of the peculiarities in the H-R diagram of a cluster can be explained in an elegantly simple fashion by the application of the results of our stellar evolution computations, which indicates that the observed properties of clusters do indeed serve to support the essential correctness of the evolutionary models for stellar development. Second, understanding the H-R diagram for clusters as a consequence of their history gives us a way to determine the age of star clusters.

Applying this technique for determining the age of clusters, we learn a bit of cosmic history. The globular clusters associated with the Milky Way galaxy all have about the same age—approximately ten to twelve billion years. Galactic clusters, by contrast, range in age from newborn to ten billion years old. Star formation activity has evidently been occurring continuously in the spiral arm regions since the birth of the galaxy, while the halo region gave birth to stars only briefly during the very early period of galactic development.

The antiquity of the cosmos is vividly revealed by star clusters, and the chronology of stellar history appears to be entirely consistent with the chronology of terrestrial phenomena. Astronomers and geologists are discovering congruent chronologies. Cosmic history displays a chronological coherence the beauty and awesomeness of which we have only barely begun to appreciate. We will explore a few additional examples in the next chapter.

CHAPTER NINE

Cosmic Chronology and Evolution

THE EVOLUTIONARY FRAMEWORK OF COSMIC HISTORY

Lessons from Stellar Evolution

Our review of the general outline of stellar evolution provides us with more than a mere list of events and processes in the temporal development of the material systems we call stars; it also provides us with strong suggestions concerning the character of cosmic history in general and the patterns for the temporal development of the entire universe. The credibility of these suggestions and the applicability of these patterns to material systems other than stars then becomes a matter of empirical testing.

What are some of the lessons that may be learned from our study of stellar evolution? What aspects of that phenomenon provide valuable clues to the way in which we might explore the behavior and history of other segments of the cosmos? We could make a rather lengthy list in response to such questions; let's look only at those items that appear particularly relevant to our larger discussion.

1. Strict Boundary Conditions Apply

The physical processes that contribute to the evolution of stars are not arbitrarily imposed; they are drawn from the ordinary patterned behavior of matter. The phenomena involved in the temporal development of stars are precisely the phenomena that constitute the normal patterns for material behavior. Models for stellar behavior were constructed under strict boundary conditions. The material of which a star is composed is required to obey the rules for motion, gravity, thermodynamics, energy conservation, nuclear interactions, and all other physical laws as far as we know them. No arbitrary new rules are introduced in the birth of stars; none of the known rules are arbitrarily suspended. Stellar evolution, if it is to occur at all, must occur within the bounds for the patterned behavior of matter that are universally exhibited.

2. *The Consequences Are Inevitable*

Given the observed properties and behavior of matter, stellar evolution is a necessary consequence. Under specifiable conditions, the gravitational collapse of a globule into a protostar will necessarily occur; there is no other option unless the law of gravity is suspended. As the protostar contracts, the conditions for thermonuclear fusion will be satisfied and fusion will necessarily occur; there is no other option unless the rules for nuclear reactions are suspended. A main-sequence star converts hydrogen into helium and forms a helium core, which transforms a main-sequence star into a red giant star. Red giants eventually lose their battle with gravity and collapse, leaving their crushed remnants in the form of white dwarfs, neutron stars, or black holes. These transformations *must* occur unless the rules for all matter everywhere are somehow suspended for stellar matter. Such an arbitrary suspension of the rules would be entirely unwarranted. Stellar evolution, then, is the necessary consequence of the ordinary patterned behavior of matter and material systems. In the context of the manner in which material behavior is governed, stellar evolution is inevitable.

3. *To Deny Stellar Evolution Is to Demand Incoherence*

The physical processes that contribute to stellar evolution are just the ordinary processes that constitute the normal patterns of material behavior. If we grant these patterns of behavior for the matter of which stars are made, stellar evolution is the inevitable consequence. To deny the reality of stellar evolution, therefore, is to deny the coherence and continuity that we observe both in our daily experience and in laboratory investigation of material behavior. To deny that coherence and continuity without warrant is arbitrarily to introduce incoherence and discontinuity into the realm of natural phenomena. Though I have encountered such denials in my reading, I find them entirely unjustified and without merit on any grounds—scientific, philosophical, or theological.[1] To deny that stellar evolution has occurred and is now occurring is to demand that the universe behave incoherently.

4. The Concept of Stellar Evolution Brings Order into a Chaos of Stellar Differences

When we first reviewed the results of observing and measuring stellar properties, we wondered at the reasons for both the similarities and the differences among stars. Why, we asked, do most stars fall into distinctly different families? How might these families be related to one another? Until the temporal development of stars was understood, these questions remained unanswered. Stars were simply different from one another and that was that. The results of stellar evolution computations, however, have solved the riddle of stellar types and introduced chronological ordering into the picture. We can now see that the various families of stars mark various stages of temporal development. The chaos of stellar differences has been removed by the discovery of chronological ordering. Each star is now better understood in the context of its historical development. Our understanding of stars has been enriched by the discovery of their evolutionary development, which supplies the reason (in the immanent and proximate sense) for both the differences and the similarities that stars exhibit. Stellar differences are not merely accidental, nor are stellar similarities merely coincidental. Both are a natural consequence of the ordinary patterned behavior of matter that makes stellar evolution inevitable.

5. Evidence Reveals Not Only Stellar Antiquity but Stellar History

In Chapter Eight we noted that the peculiar properties of stellar clusters gave us a method of determining the age of the clusters. We also noted that this method has been used to show that many clusters in the Milky Way galaxy are approximately ten billion years old. Now, some people might dismiss that information as

---

1. Most recent creationist literature is on shaky ground in this way. See, for example, Henry M. Morris's *The Stars of Heaven*, ICR Impact Series, no. 10 (San Diego: Institute for Creation Research, n.d.), Paul M. Steidl's *The Earth, the Stars, and the Bible* (Grand Rapids: Baker, 1979), and John C. Whitcomb's *The Bible and Astronomy* (Winona Lake, Ind.: BMH Books, 1984).

merely a number. Numbers by themselves are not especially interesting as a general rule, and some people may feel that such astronomical age figures are unimportant or irrelevant: "Six billion years, or six thousand years," they say; "what's the difference anyway?"

In my judgment, there is an important difference. It lies not simply in the numerical accuracy for the value of a time period but in the authenticity of the history of events that have occurred during that period of time. If someone were to claim that I am only forty-seven minutes old, rather than forty-seven years of age, that individual would be denying the authenticity of the experiences and events I believe have shaped my present character. Similarly, if someone were to claim that a star cluster is not the six billion years of age that its H-R diagram and a good deal of other physical evidence indicates, but that it is only six thousand years old, that individual would be denying the authenticity of the sequence of events and processes that have given the cluster the unique properties it exhibits. The observed properties of a star cluster are not simply evidence of their antiquity; far more importantly, they are evidence of their history. An entire history, formed by a continuity of causally related processes and events, contributes to the properties we can observe today in a cluster of stars or an individual star or a galaxy or a planet or a person.

One cannot deny the antiquity of any celestial object without at the same time denying the authenticity of the entire history of events that are revealed by that object's properties. The question is not simply whether the object is old or young; the question is whether the physical record of the object's history is authentic or merely an elaborately detailed fiction.

6. *Scientific Study Reveals Nothing about a Star's Status*

As we noted in earlier chapters, the scientific investigation of any material process—stellar evolution included—reveals nothing about the relationship of material systems to nonmaterial powers or beings. Astronomy, for example, does an excellent job of studying the "internal affairs" of stars, galaxies, and planets but contributes nothing to an understanding of any relationship they might have to any nonmaterial or spiritual being. Astronomy gives no answers to the important questions concerning the status, origin, governance, value, or purpose of the material world. Quite specifically, it provides no answers to questions that lie within the domain of the biblical teaching about the celestial luminaries as Creation. We will explore the relationship of scientific and biblical perspectives more thoroughly in Section III of this book. In the meantime, let's confine ourselves once again to the scientific purview.

Suggestions for Further Investigation

Since we have seen that the scientific investigation of the properties and behavior of stars has been fruitful and has revealed their fascinating history of temporal development—their evolutionary behavior—it is only natural to ask whether the study of other material systems would lead to similar results. If the physical properties of stars change as a consequence of material processes occurring in them, would not other material systems also exhibit an orderly, coherent, causally continuous temporal development?

The evidence gathered in search of an answer to this question appears very strongly to support the conclusion that the history of the entire cosmos is characterized by an orderly, coherent temporal development. Cosmic history is evolutionary in character. This is manifested, as we have seen, in the temporal development of stars. It is also revealed in the evolution of galaxies, though the details and mechanisms of their development are less well understood at this time. More clearly, it is manifested in the spatial and elemental evolution of the whole cosmos—topics we will discuss shortly. My colleagues in geology inform me that the evolutionary character of planetary history is firmly established. Similarly, biologists inform us that the physical record of the history of life on earth displays all the characteristics of temporal development. Theories of biological evolution are attempts to understand the mechanisms of material behavior on the biological level that contribute to this temporal progression of life-forms.

Let's briefly explore a few examples of the evolutionary character of cosmic history in material systems other than just stars.

WATCHING HISTORY HAPPEN: CONSEQUENCES OF THE FINITE SPEED OF LIGHT

The Cosmic Distance Scale

Using a variety of techniques, some geometrical, others based on the behavior of light, we can determine the distances to many celestial objects. From these measurements we learn not only about the cosmic distance scale but also about the structure of the cosmos. For the purposes of this overview, let's concentrate on three types of structures: our solar system, galaxies, and the visible universe.

Solar system is the name given to the sun and the family of lesser bodies that are held gravitationally in orbits around it—planets, asteroids, comets, and meteoroids. Planet earth is located about 93 million miles from the sun, while Pluto, the most distant of the known planets, maintains an average distance that is forty times greater, approximately four billion miles from the sun.

Other suns (the stars, that is) are located at considerably greater distances from us. To specify the value of these immense distances, it is convenient to define a length unit that is much larger than the familiar mile. The "light-year" is defined to be the distance that light travels in a year's time. Since light travels at a very high speed—186,000 miles per second—the light-year is an extremely large distance compared to the sort we deal with on a daily basis: approximately 5,880,000,000,000 miles, which is about 1,500 times the distance from the sun to Pluto. Proxima Centauri, the star nearest to the sun, is a little more than four light-years from us, which is a typical interstellar distance in our neighborhood. Sirius, the brightest star in our sky, is nearly nine light-years away; Betelgeuse, the red giant star in Orion, lies about five hundred light-years distant; and Deneb, the brightest star in the constellation Cygnus, is approximately 1,400 light-years away from us. The great interstellar nebula in Orion is located at a distance of 1,500 light-years, while the globular cluster seen in the direction of the constellation Hercules is approximately 30,000 light-years from our tiny planet circling the sun.

Thousands of nebulae and clusters, along with hundreds of billions of stars, are arranged in a spiral-structured disk we call the Milky Way galaxy. Our sun, with its tiny planets huddled close to it, is but one quite ordinary star among the billions that constitute the galaxy. We are located about 30,000 light-years from the center of our galaxy, the diameter of which is approximately 100,000 light-years. Whenever we look from earth toward the plane of our galaxy, we see the greatest number of luminous objects. This causes the "milky" streak across the nighttime sky. In other directions we see only a relatively few nearby stars against the black background of the intergalactic void.

Beyond the Milky Way galaxy lie numerous other galaxies,

most of them rather similar to our own. One of our nearest neighbors is the Andromeda galaxy, located at a distance of two million light years—a distance equal to approximately twenty times the diameter of the Milky Way. With the aid of large telescopes, countless other galaxies can be viewed. Most have dimensions on the order of 100,000 light-years; neighboring galaxies are normally separated by millions of light-years. Estimates for the number of such galaxies lie in the neighborhood of one hundred billion.

The distance to the farthest galaxies in the visible universe is incomprehensibly great. It can be stated numerically, but can hardly be imagined. Reliable methods of measuring galactic distances indicate that the visible universe—the entire cosmos within view—has a radius of more than ten billion light-years. How can we begin to comprehend such a distance? In day-to-day experience, we have gained some comprehension of distances measured in miles, tens or hundreds of miles, even thousands of miles. But light-years, thousands and millions of light-years, and surely billions of light-years are so far beyond our experience that we can scarcely conceive of their immensity. Nonetheless, measurements of the distances to remote celestial objects can be performed, and the limits of our imagination or experience ought to cast no shadow of doubt on the reliability of either the measurements or their interpretation. We do live in a cosmos of unimaginably vast dimensions.

Distance Reveals Both Antiquity and History

Light travels at a speed of 186,000 miles per second. That's fast, but not infinitely fast. Light does require a certain amount of time to travel from place to place. It takes light about eight minutes to travel the ninety-three million miles from the sun to the earth, for example. That finite travel time has an interesting consequence: when we look at the sun, we do not see it as it *is* right now, but as it *was* eight minutes ago. If we note a solar flare—a jet of incandescent gas leaping from the surface of the sun—we are in fact seeing something that took place eight minutes before. We call that eight minutes the "lookback time" of the sun as viewed from earth.

The extraordinary consequence of the finite travel time of light, then, is that when we look beyond the confines of the earth in any direction, we are in fact looking into the past! We are not able to see any astronomical events or processes as they are happening now, but only as they were happening at an earlier time. The farther an object is from us, the greater its lookback time will be, and so the farther in the past it will be from our perspective on earth. In any case, astronomical observations not only provide us with *evidence* of historical events; they actually allow us to see those past events *as they were happening* long ago—in some cases billions of years ago!

This ability to see history happening is not the result of our cleverness or our skill in devising and fabricating highly sophisticated observing instruments. It is, rather, an unavoidable consequence of the finite speed of light. We might call this ability a "gift" because it is something for which we need not work. But this gift of being able to see distant events as they were happening allows us to learn a great deal about the past—about cosmic history. We see solar events as they were happening eight minutes before the present time. Because the Orion Nebula is located 1,500 light-years from us, its lookback time is 1,500 years, and so we see it as it was 1,500 years ago. Similarly, we see the globular cluster in Hercules as it was 30,000 years ago. And we observe the Andromeda galaxy not as it is now but rather as it was two million years ago. The light we are now receiving from quasars, the most distant and luminous objects known, is many billions of years old.

Once again we see that observational evidence tells us not merely of antiquity but of history. Our view of a distant galaxy or quasar reveals not only that it has a particular age but also that it was the site of certain events millions or billions of years ago. To deny the antiquity of the cosmos, therefore, is to deny the authenticity of the history that is revealed by the information (in the form of light) that we are now receiving from distant sources.

Galactic Evolution

Since it is possible to see galaxies as they were in the past, exhibiting the properties and behavior that belong to their history, we have an opportunity to investigate galactic evolution. By comparing nearby galaxies with distant ones we may look for systematic differences related to their age. If all galaxies were formed at about the same time (and there is strong evidence to suggest that they were), then we see distant galaxies, for which the lookback time is the greatest, as they were in their youth. Nearby galaxies, with much smaller lookback times, appear to us more nearly as they are at the present. If galaxies evolve, then the more distant galaxies should appear different from the nearby galaxies. Do galaxies change as they grow older? What does the observational evidence reveal?

There is, of course, an observational problem. Distant galaxies are more difficult to observe. But in spite of that difficulty there are certain properties of galaxies that can be compared: their luminosity, the variability of their luminosity, the size of their most luminous region, and their wavelength spectrum.

Evidence gathered so far indicates that young galaxies go through a period of highly energetic, somewhat erratic, and even explosive behavior—not so different from human adolescence. A variety of energetic objects have been detected—"N galaxies," "BL Lac objects," and "Seyfert galaxies"—that appear to be young galaxies going through periods of rapid and violent change. While the character of these changes is now beginning to form a recognizable pattern, the details of the internal behavior responsible for the observed patterns are not yet fully understood.

Because galaxies are far more complex material systems than stars, it is not surprising that our understanding of their temporal development is incomplete. While the observational evidence for galactic evolution seems convincing, the theoretical explanation has not yet been formulated to the satisfaction of the scientific community. It's a situation like that of someone who's read all but the last two chapters of a good mystery novel: we have a general picture of what's happened, we have lots of clues about how it might have happened, and we've formulated several hypotheses concerning the responsible agents and their actions, but some vital information seems to be missing, or we simply have not yet been clever enough to put all the pieces of the puzzle together. Eagerly, we read on.

Thus the study of galactic evolution proceeds. The fact of their temporal development is no longer contested; only the details of the proximate cause-effect sequence remain to be discovered.

SPATIAL EVOLUTION

Galactic Redshift

Early in the twentieth century, two avenues of the investigation of galaxies came to a fruitful intersection. Measurements of the wavelength spectra of galaxies performed by such skilled observers as Vesto Slipher and Milton Humason revealed that most of them (all but those of a few local galaxies) are "redshifted"—that is, their spectra are displaced toward the longer wavelength (red) portion of the spectrum. Each redshifted spectrum displays the characteristic absorption lines of hydrogen and other familiar elements, but their wavelength values are systematically increased; the relative amount of increase is called the "redshift parameter." At the same time, Edwin Hubble devoted a great deal of time and effort to the determination of the distances to galaxies. By the late 1920s it became apparent that the results of these two lines of investigation were related. Those early measurements, now supplemented with many more, clearly revealed that the amount of redshift is directly proportional to the distance of a galaxy from the observer. The farther away a galaxy is located, the greater is the redshift of its light spectrum.[2]

The Cosmological Interpretation

The discovery of the redshift-distance relationship raised two significant questions. First, what causes the spectrum of a galaxy to be redshifted? And second, why is the redshift related to the distance?

Initially it seemed that an eminently reasonable way to interpret the redshift was to assume that it is caused by the motion of galaxies relative to the observer. A similar phenomenon was well known for the behavior of both sound and light. The perceived frequency of an approaching source of sound is higher than from the same source at rest; the perceived frequency of a receding source is lower. The classic example of this phenomenon is the change we hear in the pitch of a train whistle as it approaches and then passes us; the passengers on the train hear a single constant pitch, but the observer waiting at a crossing hears a pitch that is higher as the train approaches and lower as it travels away. The wavelength of both sound waves and light waves from an approaching source are shortened relative to the values emitted by a fixed source, while those from a receding source are lengthened. This is called the Doppler effect.

If we interpreted the galactic redshift as a Doppler shift, then we could conclude that galaxies are receding from one

2. For an extremely well-written historical account of these and later investigations, see Timothy Ferris's *The Red Limit*, 2d ed. (New York: Quill, 1983).

another at speeds proportional to the distance separating them. The conventional statement of this velocity-distance relationship is called the "Hubble law" and can be written as follows:

$$\text{recessional velocity} = H \times \text{distance},$$

where H is the "Hubble parameter." In the context of the traditional Doppler interpretation of galactic redshift, the Hubble law suggests that galaxies are separating from one another in the same manner that molecules in a sample of gas would expand into empty space.

However, while it is still commonly employed in elementary textbooks and popular literature, the traditional Doppler intepretation appears to be inadequate. Accounting for matters of motion, space, and time on a cosmic scale requires more guidance than analogy to familiar phenomena is able to provide. A thorough understanding of the implications of the redshift-distance relation requires delving into the nature of cosmological models such as those first introduced more than fifty years ago by de Sitter, Einstein, Robertson, Friedman, and Lemaitre. Such an undertaking lies well beyond the scope of this discussion; nevertheless, we would do well to note a few of the fundamental points uncovered by more sophisticated cosmological theory.[3]

It appears that galactic redshift does indeed imply motion, but not in the ordinary sense; it is not the result of the Doppler effect. The major component of the motion of a distant galaxy arises not from its movement *through* space but rather its movement *with* space. All viable cosmological models are characterized by a description of the manner in which space itself is expanding from some initially compact state. It is the spatial framework of the whole cosmos that is growing, or evolving. The galaxies are, in a sense, merely going along for the ride. They are luminous markers that reveal what the underlying framework of cosmic space is doing. Galactic redshift is the manifestation of spatial evolution. From another perspective, galactic motion might be seen as the cause of spatial expansion. The reciprocal interaction of matter and spacetime sometimes makes it difficult for us to differentiate cause and effect in this case.

---

3. For an outstanding discussion of cosmology written for the serious reader, see Edward R. Harrison's *Cosmology: The Science of the Universe* (Cambridge: Cambridge University Press, 1981); Harrison cites many other helpful references.

Though it is difficult to comprehend, we must try to visualize space itself expanding in time. Perhaps it would help to think of a deflated toy balloon covered with a series of tiny dots at regular intervals over its entire surface—the surface of the balloon representing a two-dimensional "space," and the dots representing galaxies distributed throughout that space. As the balloon is inflated and the surface stretched, the dots will all move and will become increasingly distant from one another. In much the same way, though of course in a more complex, three-dimension setting, we can imagine space marked by a kind of coordinate grid. As time passes, the grid expands, and objects (such as galaxies) fixed to it are said to be "comoving." The distance between comoving galaxies increases as space expands, each intergalactic distance stretching proportionately. Light waves emitted by one galaxy are also stretched while en route to another: this is the cosmological redshift. The greater the distance between galaxies, the more time it takes light to travel that distance, and the greater the amount of stretching that will occur. If the rate of expansion is constant, then the cosmological redshift must be directly proportional to the distance between source and observer. If the expansion rate is variable, however, the redshift-distance relationship will deviate from the direct proportionality exhibited by the Hubble law. One of the goals of observational cosmology is to see if such a deviation exists.

All modern cosmological models view the cosmological redshift as a manifestation of the amount of cosmic expansion that has occurred in the time intervening between the emission and observation of light. In that sense the cosmological redshift is the cumulative product of cosmic history between emission and observation. The light received from a remote galaxy contains information not only about events of long ago, when the light was first emitted, but also about what has happened to it since that time. Radiation received from afar reveals more than just the nature of its source; even spatial evolution leaves its mark on the spectrum of a distant source.

The Time Scale for Spatial Evolution

The differences among cosmological models currently being considered reside primarily in the manner in which the expansion rate is assumed to vary with time. The simplest model proposes expansion at a constant rate; others present the possibility of acceleration or deceleration by taking into account such forces as gravity, which tends to oppose expansion and would thus lead to a reduction in the expansion rate.

If we choose, for the sake of illustration, the simplest model, which assumes expansion at a constant rate, then we can calculate the age of the expanding universe. Rewriting the Hubble law as

$$\text{distance} = \text{recession velocity} \times 1/H,$$

we can interpret the quantity 1/H as the time during which cosmic expansion has proceeded, bringing galaxies to their present values of distance from us. Called the "Hubble period," the value of 1/H lies in the range of ten to twenty billion years. If no deceleration has occurred, the age of the universe is equal to the Hubble period; with deceleration included, it would be a bit less, but still in the fifteen-billion-year ballpark.

The size of the visible universe is limited by its age. To be visible to us, an object must lie within a distance that light could have traveled during the finite duration of cosmic history. We conclude, therefore, that we should not expect to see objects beyond a distance of approximately fifteen billion light-years. Whatever may lie beyond that horizon of visibility also lies beyond our knowledge.

These values for both the size and age of the observable universe are remarkably close to values determined by other means. Indeed, the spatial evolution of the cosmos appears coherently related to both stellar evolution and galactic evolution. The observed properties and behavior of the cosmos strongly suggest that galactic evolution has occurred within the framework of spatial evolution and that stellar evolution is occurring within the context of galactic evolution. Cosmic history appears less a quilt of random patches than a tapestry displaying a magnificent, coherent pattern—or would it be more accurate to call it a "design"?

ELEMENTAL EVOLUTION

Remnants of the Big Bang

1. Setting the Stage

The cosmological redshift reveals that the universe is expanding; galaxies comoving with the spatial framework of the cosmos are continually drifting farther apart from one another. The future of the cosmos is one of increasing galactic loneliness as our neighboring galaxies quietly retreat beyond the horizon of visibility.

But if the future is characterized by the increasing distance between galactic neighbors, the past is characterized by cosmic crowding. Contemporary cosmology attempts not only to discover the present structure and behavior of the universe on a cosmic scale; it also seeks to reconstruct its past from the multitude of clues that events and processes in its history have left behind.[4] Projecting back into cosmic history, we find that the earlier phases of our expanding universe must have been characterized by higher values for both density and temperature. Ultimately, this journey backward into history brings us to an instant at which the dimensions of the universe must have been infinitesimally small, while the density and temperature must have been infinitely large. The evidence would seem to suggest that the whole cosmos began with a vigorous burst of expansion—what has come to be popularly called the "big bang."

That burst of energy set the stage for the events and processes that followed as the constituents of the cosmos began their patterned interaction in the context of an expanding spatial framework. As if following a carefully devised script, each character has made its appearance on the cosmic stage and has begun to play its role in the unfolding drama. The appearance of the constituents of the cosmos, their interactions with one another according to set patterns, and their arrangement into progressively more intricate, complex, and functional structures forms the drama of cosmic history. From the many themes and episodes of that drama, some of which we have already considered, we ought now to take a closer look at one in particular—the theme of elemental evolution.

2. *Cosmic Background Radiation*

Following a sequence of processes and events consistent with all of the ordinary patterns for material behavior (though under extremes of circumstance far beyond those which prevail at the present time), cosmic history includes the formation of the fundamental constituents of the universe from its primordial energy. By "fundamental constituents" I mean the elementary particles, chemical elements, and electromagnetic radiation that

---

4. For an exceptionally lucid description of the "standard model" of contemporary cosmology, see Steven Weinberg's *The First Three Minutes* (New York: Basic Books, 1977). For accounts that contain a bit more technical and up-to-date information, see Joseph Silk's *The Big Bang* (San Francisco: W. H. Freeman, 1980) and James S. Trefil's *The Moment of Creation* (New York: Scribner's, 1983).

constitutes the corporeal cosmos. In their attempt to reconstruct the history of the spatial evolution of the cosmos, modern cosmologists have discovered that the temporal development of cosmic geometry and of cosmic constituency are inseparable. What happens to space necessarily affects what happens to matter and radiation, and vice versa. The coherency of material behavior binds together what might initially appear as remote phenomena. Galactic redshift has implications for the history of elementary particles, and the first moments of cosmic history have directly affected the very composition of the cosmos billions of years later.

According to contemporary cosmological models, the early and extreme-high-temperature phase of the universe was dominated by the effects of electromagnetic radiation. Initially, far more energy appeared in the form of radiation than of material particles. But after nearly a million years of expansion and cooling, the universe switched from the initial radiation-dominated era to the present matter-dominated era. At about the same time, when the temperature had fallen to about 3000°K, cosmic matter became for the first time transparent so that radiation could move freely through space. If these cosmological models are accurate, the thermal radiation released at that time should still be with us, though considerably stretched in wavelength as a result of the cosmological expansion that has occurred in the meantime.

In 1965 this "cosmic background radiation" was first detected and identified. Numerous additional measurements have confirmed that we are indeed bathed in residual thermal radiation of about 3°K. Since the temperature of the universe at the time this radiation was released into a transparent environment was 3000°K, we can conclude that the wavelength values of the background radiation have been increased by a factor of a thousand, which is remarkably consistent with the computed value of the redshift parameter that this radiation would theoretically be expected to exhibit. The evidence is quite convincing, then, that the observed 3°K background radiation was formed when the dimensions of the cosmos were one thousand times smaller than at the present time. The cosmic background radiation is a fossil remnant of the original big bang fireball, the afterglow of that primordial event.

3. *Primordial Helium*

In addition to the background radiation, there is another very important fossil remnant of primordial cosmic history. Prior to the release of the thermal background radiation, the universe was far more dense and hot. For a brief period of time, the conditions for thermonuclear fusion were satisfied throughout the universe. The first nuclei on the scene were the nuclei of hydrogen. In a sequence of reactions similar to those that were later to occur within the interiors of stars, about twenty-five percent of this primordial hydrogen fused into helium. According to computations, most (about ninety percent) of the helium present in the universe today was formed during the first few minutes of cosmic history. Soon, however, the temperature of the rapidly expanding universe dropped below that required for fusion, and so it ceased, not to resume for about a billion years, until galaxies began to assemble, providing the context in which stars could form and nucleosynthesis (the assembling of nuclei by the fusion process) could once again take place.

The current rate of helium production in stars is only about one tenth the rate required to produce the helium now observed in the universe. Very interestingly, however, helium in precisely the right amount appears as a by-product of the events occurring during the first moments of cosmic history. Is this result of cosmological theory mere coincidence? Or is it evidence for the presence of a grand design? Is it a revelation, like so many others that we have considered, that cosmic history is coherent—every element a vital link in a chain of events that begins with exnihilation and proceeds link by connecting link to the present moment?

Nucleosynthesis in Stars

Although thermonuclear fusion in stars does not satisfactorily account for the observed abundance of helium, it does provide an adequate means of producing all of the heavier elements. Stellar nucleosynthesis, particularly in the more massive stars, is a process that builds progressively heavier elements.

Earlier in our study of stellar properties we noted that the chemical composition of stars is remarkably similar. By mass, about seventy-five percent of a star's material is hydrogen, while most of the remaining twenty-five percent is helium. This is essentially the chemical composition of the material inherited from the big bang event—the material that gravitationally collected to form the galaxies within which stars are born. Stars born very early in galactic history, such as the stars found in globular clusters, began with this primordial mixture of hydrogen and helium. Even in these very old stars, elements heavier than helium—collectively called "metals" by astronomers—constitute only about one tenth of one percent of the total mass. The spectra of these stars reveal that they are quite "metal-poor."

Other populations of stars, however, have been found to have a higher percentage of metals in their chemical composition. These "metal-rich" stars are typically the younger stars found in the spiral arm regions of the galactic disk. Why should these younger stars display a higher metal content? Stellar history provides the essential clue. Any massive stars formed early in galactic history would soon complete their brief lifetimes of nucleosynthesis. Even more importantly, violent events such as supernovas provide the conditions for both the manufacture and distribution of a vast array of the heavier elements. The result of these explosions is the dumping of large amounts of material, greatly enriched in their heavy-element concentrations, back into the interstellar medium from which future generations of stars may eventually form. This recycling of stellar material leads to a progressive increase in the metal content of newborn stars. Those regions in the Milky Way galaxy most active in star formation show the greatest effects of this recycling phenomenon. Young stars in the spiral arm regions have been found to have metal concentrations as high as three to four percent.

As galaxies age, the composition of the cosmos will change as a consequence of this "elemental evolution" phenomenon. In effect, it is the price that must be paid in order to have stars powered by thermonuclear fusion. Hydrogen is consumed; heavier elements are manufactured. The light bill for the cosmos is paid in hydrogen.

PLANETARY FORMATION AND EVOLUTION

If the sun was formed by the gravitational collapse of a globule of gas and dust, how were the earth and the other planets formed? The evidence gathered so far strongly suggests that planets and other minor members of the solar system were formed as by-products of the process that led to the birth of the sun. A collapsing protostar will generally have some rotational motion. As the protostar collapses, that rotation becomes more influential in determining the outcome of the process. It appears that in most cases some material from the collapsing nebula is left in a rotating disk around the protostar. The orbiting disk of gas and dust that apparently surrounded the newborn sun is commonly called the "solar nebula."

Initially, the solar nebula would have been quite warm; computations suggest temperatures as high as 2000°K in the inner regions closest to the sun. As the nebula cooled, various materials would have precipitated as solid particles—first the compounds with the highest freezing points, then metals such as iron and nickel along with various silicate minerals. At lower temperatures, several carbonaceous materials would have condensed. Finally, as the nebula cooled below 500°K, various hydrated minerals would have formed, and, particularly in the cold outer region of the nebula, the ices of water, methane, and ammonia would have frozen out. The chemical composition of the planets in our solar system appears to vary in precisely the manner suggested by this cooling sequence. Mercury is rich in refractory materials, Venus and Earth are made mostly of metals and silicates, Mars is rich in mineral forms of iron oxides, and the outer planets are made mostly of icy, hydrogen-rich materials.

As a consequence of further gravitational interactions and collisions, the condensed particles eventually gathered into a smaller number of larger bodies, principally the major planets and their moons.[5] Once the planets were gathered into individual lumps of material orbiting the newborn sun, further processes occurring both within and on the surface of each planet continued to modify their structure and appearance. The history of such processes and events on earth is the object of geological investigation.[6]

One interesting feature of solar system history deserves further mention: the matter of what may have initiated the birth of the sun. Suppose that the globule from which the solar system formed was on the ragged edge of satisfying the conditions for gravitational collapse—a situation that may be quite common in interstellar nebulae. What event or process might have triggered the beginning of its collapse to form the solar system? An answer to that question is suggested by the confluence of two lines of investigation: the studies of interstellar nebulae and of meteorites.

In Chapter Eight we considered the properties and behavior of star clusters, briefly noting that there are several phenomena that can stimulate a burst of star-forming activity in a

---

5. My treatment of this process is necessarily very brief. For a more extended discussion, consult a text on the solar system such as William K. Hartmann's *Moons and Planets*, 2d ed. (Belmont, Cal.: Wadsworth, 1983).

6. John S. Shelton provides a good introduction to this fascinating subject in *Geology Illustrated* (San Francisco: W. H. Freeman, 1966).

large nebula. One such phenomenon is the supernova. If a massive star forms early in the sequence, it will soon complete its short lifetime of intense activity and will cap it off with a spectacular supernova. That explosion sends a shock wave through the surrounding nebula, compressing the gas in its path. Any prestellar globule close to satisfying the gravitational collapse conditions is pushed over the edge to begin its collapse and contraction into a star. If a supernova had occurred in the vicinity of the solar globule, then, we might well suppose that it could have induced the birth of the sun and its accompanying minor bodies.

But is there any evidence that such a sequence of events did occur? Here the study of meteorites makes a significant contribution to the investigation of solar system history. Meteoroids are small solid bodies, some metallic, some rocky, that occupy widely varied orbits around the sun. Occasionally one of these meteoroids, which come in sizes anywhere from that of a grain of sand to a mountain, gets on a collision course with the earth. Entering the earth's atmosphere at high speed, the meteoroid's surface is heated to incandescence by friction with air. This glowing body racing across the sky may be visible to observers as a "shooting star"—or, more technically, a meteor. Most meteors are completely vaporized in our atmosphere, but a few are large enough so that at least a part survives the trip and lands on the earth's surface. Such surviving fragments are called *meteorites.*

Of particular relevance to this discussion are those meteorites found to contain the products of certain radioactive decay processes. Iodine 129, for instance is a radioactive form of the element that is known to have a half-life of seventeen million years—that is, half of the atoms of iodine 129 in any given sample will decay into xenon 129 over a period of seventeen million years; of the iodine 129 atoms remaining, half will decay in the next seventeen million years, and so on. Measurable amounts of xenon 129, which is chemically a relatively inactive element, have been found in several meteorite samples. Because of xenon's chemical properties, we know of no other way that xenon 129 could get into the meteorites except through the process of the radioactive decay of iodine 129. But iodine 129 has a half-life of only seventeen million years, and if the nebula from which the sun formed was really billions of years old, we certainly wouldn't expect to find any significant amounts of iodine 129 left in fragments of it. The most reasonable explanation for the presence of iodine 129 in the solar nebula just prior to the gravitational collapse that led to the formation of the sun, planets, and meteoroids is that a supernova had just occurred in the neighborhood. In a single act, a supernova could accomplish the production of heavy, unstable isotopes like iodine 129, the distribution of such material into surrounding regions, and the generation of a compressional shock wave that could trigger the birth of the solar system. While the evidence for this reconstruction is largely circumstantial, the weight of the evidence and the way in which it fits coherently into the general picture of star formation strongly suggests that these are not matters of mere coincidence or accident.

In light of the manner in which the solar system appears to have been formed, the entire process of stellar evolution assumes a position of critical importance. The materials of which the earth and the other terrestrial planets were formed are relatively minor constituents of the original presolar globule. Hydrogen and helium, the major constituents, were removed from that part of the solar nebula which condensed into terrestrial planets; mostly heavier materials precipitated out in the high-temperature environment. Hydrogen-rich materials are found predominantly in the outer planets, which were formed in a much cooler region. But the pertinent question is where the heavier, "earthy" elements came from. If our understanding of cosmic and stellar history is correct, these terrestrial materials were formed by nuclear reactions that occurred in stars that completed their lives long before the sun was born. Supernova explosions and other mechanisms by which stars shed mass injected these heavier elements into the interstellar medium to be recycled—used once again in the formation of new stars. But second- and third-generation stars, enriched in the heavy elements, can also be accompanied by terrestrial planets formed as by-products of the birth of the central star. Thus the presence of planet earth and all of the elements heavier than helium that contribute to its chemical composition is dependent upon stars that lived and died early in the history of our galaxy. The very chemical elements that make up the earth and its inhabitants were formed in the thermonuclear furnaces of ancient stellar cores and in the violent supernova explosions that blasted massive stars to smithereens. Even the matter of which our bodies are made is a participant in the coherent history of cosmic phenomena. We are all made of the dust of the earth; but earthdust is stardust.

BIOLOGICAL EVOLUTION

If evolution is the theme of cosmic history, with subthemes of spatial evolution, galactic evolution, stellar evolution, elemental evolution, and planetary evolution running through its entire course, what about biological evolution? In addition to the physical evidence for the coherent temporal development of space, galaxies, stars, and planets, is there also physical evidence for the coherent temporal development of biological systems—of living organisms? Paleontology, the study of past life-forms as revealed by the fossil record, very strongly indicates an affirmative answer to that question. That differing forms of life have appeared in a very definite temporal sequence is eminently clear from the physical record, and the denial of that temporal sequence appears entirely unwarranted.

The more perplexing question, however, is whether or not the temporal sequence of life-forms revealed by the fossil record is the product of ordinary natural processes—the same patterned behavior of matter that leads inevitably to the temporal development of galaxies, stars, and planets. Various theories for the mechanism of biological evolution are currently being considered in attempts to answer that question. Personally, I see no reason, either scientific or theological, to preclude the possibility that the temporal development of life-forms follows from the properties and behavior of matter in a way that is similar to the processes that lead to the birth of planets, stars, and galaxies. I believe that the phenomenon of biological evolution, like any other material process, is the legitimate object of scientific investigation. The answer will be found by empirical study, not by philosophical or theological dictation. My guess is that a fully satisfactory description for the processes of biological evolution will eventually be worked out. I would be terribly surprised to discover that we live in a universe that is only partially coherent, a universe in which the temporal development of numerous material systems proceeds in a causally continuous manner while the history of other systems is punctuated by arbitrary, discontinuous acts unrelated to the ordinary patterned behavior of matter. (Note carefully: I am not talking about the *status* of the cosmos; I am speaking only about the properties and behavior of the universe.)

CONCLUSIONS

Whether we investigate the properties, behavior, and history of stars, of galaxies, of planets, of radiation, of atomic nuclei, or of space itself, we arrive at the same conclusion: cosmic history is evolutionary in character. The theme that permeates the history of all material systems, on both the microscopic and the macroscopic scale, is the theme of continuous, coherent temporal development over a period of approximately fifteen billion years.

The chronology of cosmic history dwarfs our sense of time. When we attempt to comprehend the age of the cosmos or try to understand what it really means to be seeing history as it was happening more than ten billion years ago, we find our sense of time to be inadequate. As in the case of the spatial vastness of the universe, so also in its temporal vastness, the cosmic magnitudes demand more than our experience prepares us to comprehend. By experience we have gained a concept of time periods measured in hours, days, and years; spans of millions and billions of years are simply beyond our scale of reckoning. But we ought never to allow the limitations of our experience to stand in the way of appreciating what can be learned from careful investigation.

The cosmos that we are able to explore is not characterized by a static, unchanging structure; rather, it manifests itself to be a system that is dynamically developing according to universal patterns of material behavior. There is indeed a majestic order discernible in the cosmos, but it is the *dynamic* order of patterned behavior, not merely the *static* order of fixed structures. In the observable universe are innumerable structures, from miniscule atomic nuclei to enormous whirling galaxies, and each of these structures is subject to change and development in response to internal or environmental stimuli. We must take steps to free ourselves from ancient and medieval concepts of a static cosmos and familiarize ourselves with the concept of a dynamically developing cosmos that modern natural science has uncovered through empirical investigation.[7]

If a principal theme of cosmic history in the material realm is evolutionary development, a closely associated leitmotif is the coherence of material behavior patterns. Cosmic history is not a syncretism of discordant phenomena but a synergism of harmonious and coherent lines of development. If the general outline of cosmic history is now correctly understood, then we must conclude that all of cosmic history is a necessary precursor to the present state of affairs. Our very presence in this form requires

---

7. For a stimulating development of this point, see N. Max Wildiers's *The Theologian and His Universe* (New York: Seabury Press, 1982).

the entire spatial and temporal vastness exhibited by the cosmos. The properties of space and time, of here and now, are fixed by the history of the entire expanding universe. The temporal development of the Milky Way galaxy has contributed vitally to the properties of the solar system. The earth and its inhabitants are made from the products of both the big bang and the nucleosynthesis that has occurred in past generations of stars. Neither we nor any other part of the cosmos could exist in our present state except in the context of the entire cosmos and its history. Events of cosmic history are coherently interrelated. The temporal development of the whole universe has contributed to our material makeup.

To deny the temporal development of the universe along the lines we have outlined thus far is to deny the authenticity of the history that is revealed by the physical record. The observed properties of the universe indicate not merely its antiquity but a full history of events and processes. To close one's eyes to that evidence or to deny the authenticity of the history revealed by that evidence is to turn one's back to the purposeful coherence that history reveals. Furthermore, to substitute in place of a coherent cosmic history a set of arbitrary, discontinuous events not causally related to one another is to demand that the universe behave incoherently.

In my judgment, however, the study of cosmic history inevitably leads us to the conclusion that the universe has experienced a purposefully directed and coherent history spanning billions of years. But that conclusion provides the occasion for further questioning. We ought not to be content in knowing only the properties, behavior, and history of the material world; we should be eager to inquire further into the meaning and significance of the cosmos. Of course such an inquiry will take us beyond the bounds of natural science, and we will have to cross that boundary into the domain of philosophical or theological or religious considerations carefully. Failure to make careful distinctions and recognize the presuppositions of one's statements can lead to confusion and misunderstanding. The source of a vast amount of confusion and misunderstanding in discussions on the relationship of natural science and biblical exegesis is the failure to distinguish between two classes of questions: those that can appropriately be addressed to the material world (the Creation) alone, and those that can appropriately be directed only to the Bible. It is to the development of that distinction that we turn next.

PART III

Integrating the Two Views

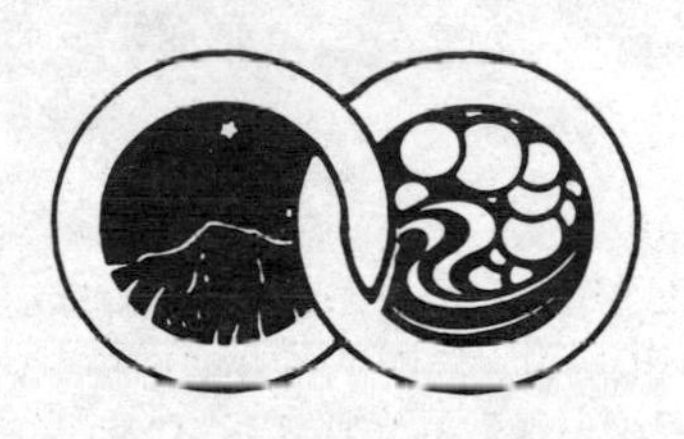

CHAPTER TEN

Taking Both the Bible and the Cosmos Seriously

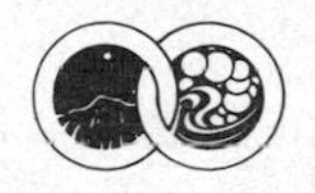

In the first section of this book we explored what it means to take the Bible seriously in its teachings about the material world. We noted that the Scriptures make a number of remarkably bold and important statements regarding the world in which we live and of which we are a part. It is from the Bible that we learn that God is the Creator, and that the entire cosmos is his Creation. After establishing that as the foundation of our view of the material world, we proceeded in the second section to explore the cosmos itself—to take it seriously in the way that natural scientists do and to discover by empirical investigation what the universe is like. Concentrating our attention on stars and other astronomical bodies, we found the cosmos to be populated with a vast array of objects that are everywhere fabricated of the same kind of substance and that this substance behaves in a coherently patterned manner that is the same throughout space and time. We also noted that the properties and behavior presently displayed by the cosmos provide abundant evidence that it has experienced a continuous and coherent history of causally related events and processes over many billions of years.

Now we should reflect on the results of these two investigations and compare the perspectives we get by viewing the cosmos alternately through the spectacles of Scripture and the lens of science. Ultimately we want to know whether these two views are contradictory or consistent. Are we faced with two disparate views that we have to choose between, or are we

The Horsehead Nebula in Orion, a dark cloud of gas and dust silhouetted against a luminous background.

presented with two complementary views that together comprise the complete picture of the cosmos in its totality?

First, however, we must establish some procedural guidelines. How do Christian natural scientists, in an effort to understand the material world, take seriously the results of the disciplined study of both the Bible and the cosmos? Specifically, what principles must guide Christians in evaluating, comparing, and synthesizing the views of the universe obtained through biblical exegesis and scientific investigation? In light of what we have considered thus far, I recommend the following as four particularly important principles: (1) we must recognize the diversity of questions that we can ask regarding the material world, and we must carefully categorize those questions; (2) we must recognize two principal sources for answers to those questions: the Bible and the Creation itself; (3) we must direct to each source only those questions that are appropriate to it; and (4) we must respect the integrity and credibility of the answers provided by each source to appropriate questions.

CATEGORIES OF QUESTIONS ABOUT THE MATERIAL WORLD

In the preceding chapters I have tried to draw attention to the importance of distinguishing among the diversity of questions that we can ask concerning the material world. Now I'd like to gather some of those considerations together and assemble a comprehensive list of categories into which all (or nearly all) of our questions about the cosmos can be placed. I find it most appropriate to divide such categories under two principal headings: *internal affairs,* which are the domain of the natural sciences, and *external relationships,* which are the domain of philosophy and theology—and thus ultimately of scriptural authority.

The Internal Affairs of the Cosmos

There are many questions pertaining to the material world irrespective of its relationship to any external nonmaterial entities, powers, or persons that might exist. These are the kinds of questions with which we dealt in Part II of this book. At the risk of oversimplification, I would like to suggest that all such questions may be placed fittingly in one of the following three categories:

1. Questions concerning *properties*. What discernible physical properties do matter and material systems possess?

2. Questions concerning *behavior*. In what manner does matter behave? What patterns of behavior do material systems exhibit? In what fashion does one material system dynamically interact with another?

3. Questions concerning *cosmic history*. What sequence of events and processes has preceded the present state of affairs in the material world? What is the character and chronology of the temporal development of the universe and of material systems within it?

Even these categories are not entirely independent of one another. Properties and behavior appear to be intimately related to one another, often by statements of immanent cause/effect relationships. Furthermore, history and temporal development can be seen as the cumulative product of the individual processes or events that constitute material behavior.

And what about human behavior? If human beings are part of the material world, do all questions of human behavior fall into these tidy categories also? No, clearly they do not. While many of our physical characteristics and behavioral patterns do follow quite directly from our material structure, humanity is not fully defined by its physical aspect alone. There exist, for example, numerous categories of human behavior in which choices and options are exercised—choices and options that we select on the basis of considerations beyond the material world. These considerations necessarily belong in the set of categories involving the relationship of the material world with nonmaterial entities.

The External Relationships of the Cosmos

In addition to having questions about the internal intelligibility of the corporeal universe, we can ask questions about the possibility and character of its relationship to external, nonmaterial entities, powers, and beings. In Chapters Three and Four, we identified a number of categories of such questions as we reflected on the biblical teaching about the cosmos as Creation. I believe that the same list of categories will serve well in this broader discussion. Though these questions have been answered in widely varying ways, they are the kinds of questions that should be considered in the context of any philosophical or religious commitment—even in the context of the view that there exist no external, transcendent beings with which the material world may be related. Let's review that list and cite a few sample questions.

1. Questions of *status*. In addition to the material world, does any order of nonmaterial entity exist? If so, where does the material world stand in relation to it? Our answer to this question will strongly influence the answers to the remaining questions in this category—questions that deal essentially with the various consequences of status.

2. Questions of *origin*. What is the ultimate cause for the existence of the cosmos? What causes something to exist in place of nothing? In other words, what is the origin of the existence of the cosmos in the past, the present, and the future? (By *origin* I mean not merely the beginning of existence but the cause for both the beginning and the continuation of existence; furthermore, I am speaking of much more than just the giving of specific form to matter that already exists, for that would be merely a matter of history or temporal development. I am using the term *origin* in the ontic sense, not simply the temporal or formal sense.)

3. Questions of *governance*. Matter and material systems behave in particular ways. What agent causes that behavior? Is the cause for material behavior internal or external to the material world?

4. Questions of *value*. Do material entities have value? If so, what criteria establish that value? Is the value inherent or does it derive from external relationships?

5. Questions of *purpose*. Does the cosmos exist for some purpose? If so, what is it? Does cosmic history display purposeful development? If so, what is the ultimate source of that purpose? To what end or goal is cosmic history moving?

DIRECTING THE QUESTIONS TO APPROPRIATE SOURCES

When we're searching for an answer to a question, as when we're searching for something that's lost, our first task should be to identify the area in which it's most likely to be found. Careful thinking and common sense are invaluable aids in both searches. We have brought our questions concerning the cosmos to two sources of knowledge: Scripture and Creation itself. Those of us who want to take the Bible seriously are already aware that the

Scriptures present the answers to many important questions about the status, origin, governance, value, and purpose of the universe. Similarly, those of us who want to take the Creation seriously are already aware that honest and competent empirical investigation of the cosmos will provide answers to questions about its physical properties, material behavior, and temporal development. Together, we have already asked many such questions and found many answers. But we have been successful principally because we have been careful to look in the right places for the answers to our questions.

The natural sciences function within a limited domain. I am convinced that the only kinds of questions about the physical world that can be legitimately directed to the Creation itself and investigated by the tools of natural science are those questions that fall within the category of the internal intelligibility of the universe—questions regarding properties, behavior, and cosmic history. Natural science is powerless to deal with any other categories of questions, and competent, honest scientific investigation is carried out in the awareness of that limitation—a limitation that all of us, regardless of profession or religious conviction, must come to recognize. Thus, when scientists make statements or conjectures concerning matters of the status, origin, governance, value, or purpose of the cosmos, they are necessarily stepping outside the bounds of scientific investigation and drawing from their religious or philosophical perspectives. Of course scientists have every right to do that, provided they do it openly and with integrity. The failure of some scientists to recognize the limits of the scientific domain and to acknowledge when they shift from the domain of natural science to matters of religious commitment has, I believe, contributed greatly to the fog that enshrouds the contemporary creation/evolution debate.

But the boundaries of the domain of scriptural exegesis must also be honored; not all questions can be appropriately addressed to the Bible for answers. I am convinced that it is both appropriate and mandatory for us to address to the Bible questions about the relationship of the physical world to God, but I consider it wholly inappropriate to address questions about the physical properties, material behavior, and temporal development of the world—questions raised in the context of the modern scientific era—to the Bible. Written with a pointed emphasis on covenantal relationships, the Bible shows little interest in questions restricted to the internal structures and physical mechanisms of the universe conceived as a material system, and competent, honest biblical exegesis is carried out in the awareness of the Bible's divinely directed focus of attention. Whether we are scientists or theologians, we are obligated to honor the integrity of Scripture's special function as the covenantal canon that reveals the unique relationship of the whole universe to its Creator-Redeemer. Thus, when theologians make statements or conjectures about geological processes or thermodynamic phenomena or cosmic chronology, they are necessarily stepping outside the bounds of scriptural exegesis and into the domain of modern natural science. Of course theologians have every right to do this, so long as they do it openly and with integrity. The failure of some biblical scholars to recognize the special focus of Scripture and to acknowledge when they shift from the domain of biblical exegesis to matters of scientific investigation has, I believe, also served to perpetuate the misconceptions embodied in the contemporary creation/evolution debate.

RESPECTING THE ANSWERS

The Answers Learned from the Study of Scripture

In answer to the fundamental question of the status of the material world, the Bible is both clear and forceful. In sharp contrast to various forms of ancient paganism, the Bible boldly pro-

Categories of questions about the material world	Appropriate sources of answers for the Christian
A. Internal affairs 1. Properties 2. Behavior 3. History	The created cosmos itself, which is constituted and governed in such a way that it is amenable to empirical investigation and is intelligible to the human mind.
B. External relationships 1. Status 2. Origin 3. Governance 4. Value 5. Purpose	The Bible, the covenantal canon, which was written principally for the purposes of revealing the divinely established covenantal relationship among God, mankind, and the rest of Creation, and of providing a witness of past human experience with the Creator-Redeemer.

TABLE 10-1. ESSENTIAL GUIDELINES FOR TAKING BOTH THE BIBLE AND CREATION SERIOUSLY. If we address to the Bible and to the Creation only appropriate questions, no conflicts or contradictions will ever arise.

nounces that the whole cosmos—the heavens, the earth, every living thing—is God's Creation. It does not have the status of deity; it stands separate from the one God, beneath him as his covenantally bound servant. Genesis 1, functioning as preamble to the covenantal canon, introduces God as the Creator and establishes the status of all else as Creation.

As Creation, the cosmos owes its entire existence to God—not only its temporal beginning, not only its structural form, but its very existence at each moment in time. This is to say that the cosmos is radically contingent; without the preserving action of the Creator, nothing would exist. "It is in him that we live, and move, and exist" (Acts 17:28).

As Creation, the cosmos is governed by the Creator. The material behavior of the cosmos may be related to its physical properties by immanent cause-effect relationships, but the transcendent cause of all action, even of the cause-effect relationships, is the effective will of the Creator. Without the continuing action of the Creator, no material behavior would be possible.

As Creation, the cosmos and every creature in it has value. The value is not inherent in material properties or behavior; it lies in the relationship between Creation and Creator.

As Creation, the cosmos and all of its history displays purpose. Cosmic history is being directed by the Creator. All temporal development within the universe reveals the Creator's intentions; cosmic history represents the unfolding of the purposes of the Creator for his Creation.

In the context of this biblically revealed relationship of the Creation to its Creator, we can draw certain inferences concerning the general character of the material world and its behavior. In this way, the Bible's revelations concerning the world's status and the consequences of that status serve as "control beliefs" that govern our expectations concerning what types of properties, behavior, and history the cosmos might exhibit.[1] For instance, because the cosmos is God's Creation, we expect it to exhibit an orderly structure and to behave according to consistent, coherent patterns. Because God has made us in his image and directed us to be responsible stewards of Creation, we expect that he will have made the material world such that we

---

1. For a challenging discussion of the relationship between religious commitment and scientific theorizing, see Nicholas Wolterstorff's *Reason within the Bounds of Religion* (Grand Rapids: Eerdmans, 1976).

will be able to understand and work with it as we should. We expect it to exhibit proximate causality, to display correlations between properties and behavior that can be empirically determined. We expect to live in an environment in which regulated and responsible behavior is both possible and required. Because cosmic history reveals the Creator's intentions, we expect to find purposefully directed temporal development toward a goal. And because the cosmos is God's Creation, it is worthy of our most earnest and careful study. The products of that study must bring praise to the Creator and they must promote the effective employment of the Creation in service to him.

The Answers Learned through the Scientific Study of the Creation

Many scholars maintain that the Christian belief that the cosmos is God's Creation, and the expectations concerning the character of its behavior that we base on that belief—particularly as it has been revitalized in the Protestant Reformation—laid the groundwork for the development of modern science.[2] As Henry Stob has expressed it, "it was Christianity that supplied the firm foundation for modern natural science, and . . . the Reformation was used by God so to delineate this foundation as to dipose men to build on it the vast new structure of science."[3] Therefore, we ought never to view natural science as the enemy of belief in the biblical teaching concerning creation; it is, to the contrary, the offspring of that belief. Natural science, the intellectual child of creation doctrine, needs parental nurture and discipline from time to time as much as any human child. And, as in the case of a human child, if we find natural science to be imperfect or limited in its capabilities, we should not for that reason ignore it or toss it out.

It is when we view the world as Creation—the handiwork of a Creator who is its Originator, Preserver, Governor, and Provider—that we perceive it as a fitting object for empirical scientific study. When we break away from viewing the

2. For varying perspectives on the role of Christian belief in the development of modern science, see R. Hooykaas's *Religion and the Rise of Modern Science* (Grand Rapids: Eerdmans, 1972), Eugene M. Klaaren's *Religious Origins of Modern Science* (Grand Rapids: Eerdmans, 1977), and Stanley L. Jaki's *The Road of Science and the Ways of God* (Chicago: University of Chicago Press, 1978).

3. Stob, *Theological Reflections* (Grand Rapids: Eerdmans, 1981), p. 4.

cosmos as a capricious or evil sentient being or as the imperfect shadow of some ethereal ideal world, we are freed to perceive it as being faithfully governed by a rational and benevolent Creator, and we can be assured that the properties of such a world will be both intelligible and meaningful, that its behavior will be coherently patterned, that its history and temporal development will display both coherence and purpose. It is for all of these reasons that Christians desire to take the cosmos seriously and to pay serious attention to the results of scientific investigation.

Motivated by a desire to know more about the celestial luminaries that are a part of the Creation, we embarked on a scientific study of the properties, behavior, and history of stars. We learned that from starlight we can determine such stellar properties as luminosity, temperature, size, and chemical composition. Furthermore, after investigating the behavior of stars, we concluded that stellar properties and behavior are fully consistent with all of the known patterns of physical phenomena. Now, because we believe that stars are a meaningful part of God's Creation, we trust that the information concerning stellar properties and behavior conveyed by starlight is authentic—that stars with these physical properties and behavior patterns really exist rather than being mere illusions or superficial appearances. Finally, by computing the sequence of changes that must occur as a consequence of the physical processes that take place within a star, we were able to determine what the life history of a star, or even a cluster of stars, must be like. Again, because we believe that stars are an integral part of the Creation and its history, we have good reason to believe that the history of stars and clusters of stars is authentic history; we believe that the processes and events indicated by their present properties actually occurred and constitute important episodes in the coherent and purposeful history of the Creation.

As Christians, then, we have a responsibility to give serious consideration to the valid results of competent scientific investigation. They represent the products of a study of the Creation—not merely of the material world, but of the Creation. It is the Creator who chose to constitute it, govern it, and guide its history in the manner we can empirically discover. We ought never to impose on it the limitations of our experience with space or time or restrict its grandeur to the confines of our limited insight. Instead, we must humbly but eagerly try to discover what the Creation is like. The natural sciences will serve extremely well in that enterprise. While their domain is limited to the internal affairs of the material world, the sciences deserve our respect, encouragement, and involvement within that domain. Because natural science deals only with matters of internal intelligibility, its story of the cosmos will necessarily be incomplete; but within the limits of that category, it has every possibility of being valid.

If the natural sciences are pursued in the awareness of their categorical limitations, they need not be feared; the results of careful and honest investigation of the material world cannot possibly lead the Christian astray. On the contrary, a richer knowledge of the properties, behavior, and history of the Creation ought to lead the Christian to a richer appreciation of the Creator and of all that he has chosen to give us in order that we might have life—both temporal and eternal, both material and spiritual.

Yet many voices continue to this day to preach the "gospel of either/or-manship." They insist that the results of modern natural science stand in opposition to the Christian faith. The propagators of this "gospel" come from a broad spectrum of religious commitments, with worldviews as different as fundamentalist Christianity and atheistic naturalism.[4] Though they differ radically in their conclusions, they appear to agree on a principal premise: that one must choose between the two views of the cosmos. I am, however, convinced that taking both the Bible and the cosmos seriously reveals that there is no contradiction—or, to put it even more strongly, that a genuine contradiction is not even possible.

THE NONCONTRADICTORY CHARACTER OF ANSWERS TO CATEGORICALLY COMPLEMENTARY QUESTIONS

Toward a Definition of Categorical Complementarity

The view of the cosmos we get when we look through the lens of science is different from the view we get when we look through the spectacles of Scripture. There's no reason to deny

4. For a sample of the fundamentalist version of either/or-manship, see Duane T. Gish's article "It is Either 'In the Beginning, God'—or . . . Hydrogen,'" *Christianity Today*, 8 October 1982, pp. 28-33, or Henry M. Morris's *Biblical Cosmology and Modern Science* (Grand Rapids: Baker, 1970). A representative naturalistic version would be P. W. Atkins's *The Creation* (San Francisco: W. H. Freeman, 1981).

that difference, and every reason to respect and appreciate it. But why do these two views differ from one another? Is it because one is valid and the other is not? Do they differ because they are contradictory? Many Christians, including the vocal spokespersons of the "creation science" movement, believe this to be the case. In my judgment, however, the two views differ not because they contradict one another but because each is incomplete: neither view by itself will give us a complete picture of the physical world. If that is the case (and I am convinced that it is), two questions naturally arise: what is the nature of the incompleteness in each view, and how are the two views related?

One approach assumes that each view provides partial answers to a common set of questions. If this were the case, nearly every imaginable question about the cosmos would have to be addressed to both Scripture and Creation. The complete (or at least the best available) answer to each question would then be formulated by piecing together the partial answers provided by empirical investigation and biblical exegesis. Each partial answer would be supplemented by another partial answer. According to this perspective, which I'll call the "supplementarist approach," both the Bible and science are incomplete in the sense that the answers each provides are full of gaps that have to be filled by the other.

One example of the supplementarist approach is traditional *concordism*, a perspective that was especially popular during the nineteenth century. In a nutshell, concordists assume that though the biblically and empirically derived views may each be incomplete in the supplementarist manner, they are still concordant—that is, their partial answers are in fundamental agreement with one another. A concordist approach to geological history, for instance, would proceed on the assumption that the early chapters of Genesis provide data from which the major features of earth's history can be determined and that this biblically derived scenario of terrestrial history would fit harmoniously with an empirically derived reconstruction of geological history. Assuming that Genesis addresses the same kinds of questions that historical geology does, a concordist might interpret the days of Genesis 1 as lengthy geological ages in order to get the time scales of Genesis and geology to harmonize.[5]

---

5. For a brief review of such attempts at harmonization, see Davis A. Young's *Christianity and the Age of the Earth* (Grand Rapids: Zondervan, 1982), pp. 55-59.

But the supplementarist approach is, I believe, doomed to failure. In bringing questions from all categories to both the Bible and the Creation, it fails to respect the differences in their characters and the consequent differences in the categories of questions we can appropriately address to each. Following this approach, for example, would lead one to bring questions of material behavior to the Bible and questions of divine governance to the physical universe, and also to bring technical questions about cosmic chronology to the Scriptures and teleological questions about transcendent purpose to the corporeal cosmos. But these combinations of questions and sources will not produce any sensible answers. Bringing inappropriate questions to either source will generate only confusion and chaos. Inappropriate questions are the seeds of nonsense—seeds that have the potential of producing successive generations of progressively more ludicrous questions and meaningless answers.

The supplementarist approach fails, then, leaving us with the question of what to put in its place. The proponents of either/or-manship insist that scriptural exegesis and contemporary science present us with contradictory views, and they demand that we choose either one view or the other. But as I have said, I am convinced that the scientific and biblical views are not at all contradictory. They are *different,* surely, but difference does not automatically entail *contradiction.* If I were to hold up a sheet of paper and one person were to describe it as square while another said it was green, they would not be contradicting each other; assuming both descriptions to be accurate, the two individuals would in fact be providing *complementary* descriptions—each describing the same object but providing answers to entirely different questions about that object. The synthesis of these two descriptions into a coherent unity illustrates several aspects of the concept of "categorical complementarity," a rather stiff title for something that we encounter daily and treat as a matter of common sense.[6]

---

6. What I am calling "categorical complementarity" is closely related to something D. M. MacKay calls "hierarchic complementarity." For MacKay's development of that concept, see his article "'Complementarity' in Scientific and Theological Thinking," *Zygon* 9 (September 1974): 225-44. In assigning a different name to this approach, I wish to downplay MacKay's emphasis on the logical character of the relationship between views and to avoid the relative ranking that the word *hierarchic* suggests. What distinguishes the biblical and scientific categories is not so much the logical level at which they are pitched as the qualities they are designed to consider.

When we investigate anything from the cosmos to a sheet of paper, there are any number of categories of questions we can ask about it. The people who looked at the sheet of paper analyzed it in terms of the categories of shape and color. A more complete analysis would have to involve categories for all qualities that are distinct from one another and yet are integral aspects of the complete entity. For the purposes of this discussion we'll call these "categorically complementary" qualities. Thinking in terms of categorical complementarity helps us, I believe, to avoid the pitfalls of both the supplementarist approach and the claims of either/or-manship. We recognize that two different answers (or sets of answers, or views) need not be contradictory if they represent replies to questions in different categories. Considering the example of the sheet of paper once again, it is obvious that the descriptions "green" and "square," while very different from one another, are not contradictory. Without the slightest fear of contradiction, we could describe the piece of paper as both green and square. If anyone would demand that we choose either one description or the other, we could quickly dismiss the demand as nonsensical either/or-manship. Furthermore, it should also be clear that each description is complete within its own category and has no deficiencies that need to be supplemented by the other: "green" is an adequate answer to the question of color, and "square" is a complete description of shape. The two descriptions, then, are neither contradictory nor merely supplementary; they are complementary descriptions that answer questions drawn from entirely different categories.

So it is, I believe, with our two views of the cosmos. The view through the spectacles of scriptural exegesis reveals information about the relationship of the cosmos to God, providing answers to questions in the categories of status, origin, governance, value, and purpose. The view through the lens of natural science, on the other hand, reveals information about the internal intelligibility of the cosmos, providing answers to questions in the categories of physical properties, behavior, and history. As long as these categorical distinctions are honored, no contradictions will arise. As a piece of paper can be at once both green and square, the universe can be at once both God's Creation and a system of material objects presenting empirically observable properties, rigorously patterned behavior, and a coherent temporal development. The two views, one pertaining to internal affairs, the other specifying the character of external relationships, are neither inherently contradictory nor merely supplementary but are related as categorically complementary descriptions. Each view reveals different qualities of the whole. Both views are necessary in order to perceive the complete unity, the created cosmos, the cosmic Creation.

It is my earnest hope that the approach of categorical complementarity will help to resolve the innumerable misunderstandings that exist concerning the relationship between the biblical and empirical views of the cosmos, its status, and its temporal development. This approach represents an attempt to take both the Bible and the Creation seriously in a manner that respects the character of both while directing praise to the Creator of all.

A Critical Evaluation of Categorical Complementarity

Though I have made some bold claims regarding the merits of the concept of categorical complementarity, I don't mean to give the impression that it's beyond criticism. It is only by critical evaluation that we can test whether it does in fact comport with biblical teaching and faithfully reflect the manner in which contemporary natural science is actally practiced. Let's consider a few challenges that categorical complementarity must meet.

Question 1: Are the internal affairs and external relationships of the cosmos really as independent and separable as the shape and color of a piece of paper? If not, how can we justify directing the two categories of questions to different sources for answers?

Independent? Not entirely. As we noted earlier, modern empirical science arose historically in the context of, and as an outgrowth of, the biblical teaching that the cosmos is God's Creation. Furthermore, we noted that it is because we view the cosmos as Creation that we can rely on its being governed in an orderly, coherent, intelligible, and purposeful manner. Thus we can see that the character of the modern scientific enterprise has already been strongly influenced by the biblical view of the status of the universe. Hooykaas makes essentially the same point in *The Christian Approach in Teaching Science:* "If . . . we were to try to build a 'Christian' science, we should be acting like a man who hunts for his spectacles while they are on his nose. Modern science and technology to a great extent *are* fruits of Christianity."[7] In light of that historical connection, we should

7. Hooykaas, *The Christian Approach in Teaching Science* (London: Tyndale, 1960), p. 12.

not argue that internal and external matters are entirely unrelated to one another.

But neither are they so closely related that the scientific investigation of physical properties, behavior, and history would also yield unambiguous answers to questions of external relationships. After the scientific emphasis on the regularity of material behavior was first established (as an outgrowth of the biblical doctrine of creation), some natural philosophers began to claim that such regularity arose not as the result of divine governance but as the product of self-governance.[8] Naturalism, based on materialistic monism (the belief that matter is the only form of being), assumes that the cosmos is essentially an independent, autonomous machine, that its existence depends on no power or being beyond itself, that its behavior is governed by the essential nature of matter itself. Proponents of naturalism maintain that the observed properties, behavior, and history of the cosmos can be correctly understood only in the context of an entirely atheistic view of its status.

It has been my argument all along that this view is simply incorrect, that a scientific study of the internal affairs of the cosmos alone will never provide definitive answers to questions of its status, origin, governance, value, or purpose. If they are to be settled at all, such matters will have to be settled beyond the boundaries of natural science. Biblical theism and atheistic naturalism each presents a claim that the observed properties, behavior, and history of the cosmos can best be understood within the framework of their particular view of the status of the material world. Personally, I am wholly committed to the belief that the biblical view is the correct one. But the warrants for that belief are not drawn from scientific investigation; instead, the ultimate warrant for that belief is the testimony of the Holy Spirit that the Bible is trustworthy in its revelation that God is the Creator and that the entire cosmos is his Creation.

Looking at the relationship between external and internal matters from the other direction, we must also recognize that perceiving the cosmos as Creation does not in itself provide the means to a complete understanding of its physical properties, its

---

8. It is reported, for example, that when Pierre Simon LaPlace (1749-1827) was asked by Napoleon why his book on astronomy failed to mention the Creator, LaPlace haughtily replied, "Sire, I have no need of any such hypothesis."

Paul Davies provides an informative and provocative discussion of the question of divine action in the context of contemporary physics and cosmology in his book *God and the New Physics* (New York: Simon & Schuster, 1983).

patterns of material behavior, or the history of its temporal development. In contending that we can have certain reasonable expectations about the cosmos because it is Creation, I am not claiming that the Bible provides us with the kind of data that is useful for evaluating specific scientific theories, nor am I suggesting that we have any grounds for demanding that biblical exegesis provide us with specific answers to questions in the categories of the properties, behavior, and history of the cosmos. Empirical investigation alone will reveal the answers to these questions. Neither biblical theism nor atheistic naturalism can justifiably make a priori demands upon the results of empirical science.

The heart of the matter is this: while the categories of internal intelligibility and external relationship are not wholly independent of one another, neither are they so closely related that the answers for one category of questions unequivocally determine the answers to the other. Each set of questions must be investigated on its own terms. Questions regarding the status of the universe and the consequences of status for origin, governance, value, and purpose cannot be answered on the basis of even the most competent scientific investigation. It's not a matter of competence but of domain. Neither can questions concerning the specific patterns of physical behavior or the chronology of cosmic history be derived from the results of the most competent biblical exegesis. If we truly wish to take both the Bible and the Creation seriously, we must honor the respective domains of biblical exegesis and natural science. I remain convinced of the soundness of my recommendation that questions of external relationship should be addressed only to the Bible and that questions of internal intelligibility should be addressed only to the Creation itself.

Question 2: Is natural science religiously neutral?

By claiming that natural science is powerless to provide answers to questions regarding status, origin, governance, value, and purpose, and by accepting the results of the scientific investigation of material properties, behavior, and history, are we thereby accepting the claim that natural science is religiously neutral? I judge that in a carefully delimited sense we are, and that we are warranted in doing so—so long as we are dealing with natural science as we have defined it and so long as it is carried out solely in the domain to which we have assigned it.

I am not claiming, of course, that natural scientists as persons are religiously neutral. No person's conception of the cosmos *in its totality* will be religiously or philosophically neutral. To perceive the cosmos in its totality requires a commitment to certain answers to the questions of status and its consequences. Answers to questions in these categories are directly affected by one's religious faith, and those who take the Bible seriously believe that only it provides authentic answers. Clearly, then, when scientists are functioning as complete persons who wish to perceive the nature and the meaning of the universe in its totality, they cannot maintain religious neutrality.

But when a scientist is performing the technical tasks of investigating only the internal affairs of the material world, is it then possible to operate in a religiously neutral manner? Or are even technical operations such as observation, measurement, data analysis, and theoretical modeling affected by the scientist's religious faith? I am convinced that at this time in the history of science they need not and ought not be affected. As long as investigators honor the categorical boundaries that we have outlined and keep their professional activity within the domain of internal affairs alone, religious bias has essentially no room to operate. Natural science performed in the manner that we have described can be as religiously neutral as dialing a telephone.

Perhaps religious neutrality is possible in principle, but is natural science actually carried out in that manner? How well do natural scientists stay within their domain of operation? My personal judgment, based on more than twenty years of experience, is that the vast majority of professional scientific work is performed and reported in a manner that honors the categorical boundaries and stays within the legitimate domain of natural science. The best illustrations of this would be found in the professional journals, which publish the majority of the technical literature of science. One could, for example, read the *Astrophysical Journal* for days on end without detecting the religious commitments of any authors, many of whom are Christians. Textbooks do a reasonably good job of maintaining religious neutrality, although one can find numerous examples of statements, usually in the introductory sections, that reveal a bias—most frequently in the direction of naturalism. But the bulk of today's textbooks provide little cause for concern.

Popular literature is another ball game. When scientists write for the general public, they usually choose to speak as complete persons, and therefore not only report the technical results of natural science but also seek to place those scientific results in the framework of their worldview. This book is one example; I very purposely chose to consider the results of natural science in the context of my Christian faith. Carl Sagan's very popular TV series and book *Cosmos* is another example. Its opening line—"The cosmos is all that is or ever was or ever will be"—reveals that Sagan wishes to place the results of natural science in the framework of naturalism. Even more troublesome, its general tone seems designed to give his audience the impression that the results of natural science provide the warrant for his anti-theistic philosophical position. Perhaps a good dose of categorical complementarity should be prescribed.

What about the material that we considered in Chapters Seven to Nine? Has that been influenced by the religious bias of certain influential astronomers? No, I sincerely believe that it has not. The numerous astronomers and physicists whose work has contributed to our understanding of stars represent a full spectrum of religious beliefs. Though their spiritual commitments are very diverse, their scientific work forms a single unit—not because they have violated or compromised their beliefs but because the nature of their professional scientific task requires that their inquiry into the internal affairs of the cosmos be distinguished and separated from their views on its status.

During the past two decades there has been much discussion on the extent and the character of philosophical, cultural, and religious influences on the scientific enterprise.[9] I am well aware that our brief survey here does not exhaust this complex topic. In this context, I simply want to stress the following points. (1) I have been describing a somewhat idealized form of natural science. While professional natural scientists may strive to carry out a program of empirical research and theoretical modeling in this idealized manner, restricting their professional concern to the domain of internal intelligibility exclusively, we can assume that none will achieve this goal completely. (2) Although the actual practice of natural science is imperfect in this way, the scientific community does have exceptionally high standards of competence and professional integrity, and its findings deserve a good deal of respect. (3) Those critics who encourage us to believe that the scientific enterprise is so biased toward a naturalistic worldview as to justify the wholesale rejection of the contemporary understanding of cosmic structure, behavior,

---

9. For an informative and highly readable discussion of this topic, see Colin A. Russell's *Cross-Currents: Interactions between Science and Faith* (Grand Rapids: Eerdmans, 1985).

and history are, I believe, seriously mistaken. Though their motivation may be admirable, their misperceptions of professional natural science are no help to the Christian community.

But should the Christian community rest comfortably with the scientific conclusion that cosmic history is evolutionary in character? Are not creation and evolution antithetical and contradictory concepts? These are important and emotion-laden questions. They deserve more than brief comment, and so we will return to them at length in Chapter Eleven.

Question 3: Does the concept of the religious neutrality of natural science violate the biblical principle that everything we do must be done in the awareness of the Creator's presence and under the lordship of Christ?

I received my training and nurturing in the Christian faith in the context of a Calvinistic heritage. One of the important themes of that heritage is that the entirety of one's life is to be lived to the glory of God and in the awareness of his presence. Our devotional, recreational, vocational, and intellectual activities, along with all other facets of daily living, must be guided by that principle. Even our professional activities as natural scientists, therefore, must be brought under Christ's lordship and done for the purposes of serving and glorifying our Creator-Redeemer.

The Dutch Calvinist Abraham Kuyper, in his *Principles of Sacred Theology*, went to some lengths to develop the "two-science" model of human thought.[10] For the Kuyperians in my audience, I'd like to make just a few remarks that may facilitate a comparison of my approach with Kuyper's.

According to Kuyper, we must divide humanity into two camps: regenerate persons, who believe in God and in such fundamental concepts as creation, divine revelation, and the need for salvation; and unregenerates, who reject God and embrace the basic tenets of naturalism. Consequently, says Kuyper, there are two sciences: "regenerate science" and "unregenerate science." One is theistic and the other is atheistic. Starting from two radically opposed positions, the two systems of thought are antithetical and irreconcilable. There is no religious neutrality in these two sciences.

At first glance, it may appear that my claim for the religious neutrality of professional natural science stands at odds with my

---

10. See Abraham Kuyper's *Principles of Sacred Theology* (1898; rpt. Grand Rapids: Baker, 1980), pp. 150-82.

Calvinistic and Kuyperian heritage; but the contrary is true. The apparent differences lie principally in terminology rather than in substance. Kuyper uses the term *science* in a far broader sense than our common use of the term denotes. In order better to understand Kuyper's discussion, we must substitute for his word *science* the term *worldview,* meaning a complete system of thought that seeks answers to questions in all categories. Thus Kuyper is pointing out to us the antithesis between Christian and naturalistic *worldviews.* Furthermore, when Kuyper denies the religious neutrality of "natural science," he is using a definition of natural science that differs significantly from mine. The term as he uses it means "the science of nature"—that is, a system of thought that seeks to answer questions about nature *in its totality,* that incorporates both our internal and external categories. The activity that I have identified with the term *natural science* Kuyper refers to as the "exact sciences," which he held in rather low esteem for being incomplete and for failing to deal with the really important matters—matters that we have placed in the category of "status and its consequences."

For the purposes of this and subsequent discussions, let's now define two additional terms that may serve to clarify a number of important issues. Recall that by our definition, "natural science" is the restricted study of the material world that deals with internal affairs alone. It will be convenient also to have identifying names for the broader conceptions of the cosmos that result from incorporating natural science into the framework of a complete worldview, drawing from that worldview the answers to questions concerning the external relationships of the universe—questions of status and its consequences. Since we are considering only two worldviews—the Christian and the naturalistic—we need name only two different sciences (in Kuyper's broad sense) of the world *in its totality* —"naturalistic science" and "creationomic science."[11]

---

11. The choice of this peculiar name requires a brief explanation. For the purposes of parallelism, the name "creationistic science" would have been a logical choice. Similarly, one might have chosen to use the name "creationism" for the worldview to be contrasted with naturalism. Both of these terms, however, have certain connotations that I wish to avoid. "Creationism" and "creationistic science" have come to be identified with a particular position in the creation/evolution debate. They not only represent a commitment to the belief that the cosmos is God's Creation but also entail a commitment to a particular concept of the manner and chronology of divine creative action (the instantaneous appearance of mature forms by divine fiat during six twenty-four-

Naturalistic science is natural science placed into the framework of a naturalistic worldview—the system of thought based on the presupposition that only the material world is real, that the cosmos is self-existent and self-governing, and that all phenomena occur as a consequence of powers inherent in matter rather than as the consequence of any purposeful, external, nonmaterial influence.

Creationomic science is natural science placed into the framework of the biblical worldview—the system of thought based on the revelation that the material world is God's Creation, that the Creation is dependent on God both for its existence and for its lawfully governed behavior, and that all phenomena occur under divine direction toward the goal of redemption in Christ.

Naturalistic science and creationomic science are clearly antithetical. Once again we see two "sciences" (in the broad sense) that are irreconcilable because of their deliberate lack of religious neutrality. Their differences, however, lie not in the matter of internal affairs—the domain of natural science—but rather in matters of external relationships—the domain of metaphysics and theology. When Kuyper speaks of the nonneutrality of science he is speaking primarily, if not exclusively, of such differences as are exhibited by naturalistic science and creationomic science—the sciences of nature in its totality. Kuyper readily admitted that the "exact sciences" (roughly equivalent to our definition of the natural sciences) are held in common by both worldviews and that they provide information (which, in my view, is restricted to the categories of properties, behavior, and history) that must be coherently incorporated into each system of thought.

Once this matter of terminology is cleared up, it should be evident that our claims for the religious neutrality of natural science (restricted sense) and for the legitimacy of the categorical complementarity approach are not at all in conflict with the basic premises of Kuyper's two-science model. Kuyper's analysis is similar to ours in that it seeks to recognize both the common

---

hour periods a few thousand years ago). To avoid identification with these additional stipulations, we must use different terms.

Furthermore, I wish to have a term that not only identifies the doctrine of creation as foundational but also emphasizes that God governs his Creation in a lawful manner. The suffix *-nomic* is derived from the Greek word for "law," *nomos,* and is included as a reminder that it is by God's law that the universe exists and behaves as it does.

features and the differences between Christian and naturalistic concepts of the cosmos in its totality. While these two concepts must hold the legitimate results of empirically based natural science in common (neither wishes to deny what can be observed concerning the properties, behavior, or history of the cosmos), they draw from antithetical worldviews their diametrically opposed perspectives on the status of the universe and the consequences of that status for questions of origin, governance, value, and purpose. Only one natural science, but two cosmic perspectives.

Before closing this discussion, we have to address one final question: Can any human activity be considered religiously neutral? My answer to this question is a qualified Yes. Why qualified? Because we must be very careful to specify which particular aspects of any given human activity can be included in our claim of religious neutrality.

Any human act *in its totality* will be influenced by one's religious perspective. In its totality, any human act will, for example, be motivated by some desire, directed toward some goal, and performed in some particular attitude. Those individuals who place all of their activities under the lordship of Christ will necessarily allow their purposes and attitudes to be fully conformed to his will. Those individuals seeking to operate in the context of some other religious or philosophical commitment will have discernibly different purposes and attitudes.

There are, however, selected aspects of many human activities for which the lordship of Christ makes no perceptible difference—baking a cake, for instance. If a Christian and a non-Christian were to set themselves to the task of baking a particular kind of cake, each would gather the same ingredients, use the same kinds of utensils and appliances, and follow the same procedures. All things being equal, the cakes would be essentially identical; like the loaves of bread we buy in the grocery store, the cakes would carry no clues to reveal the religious commitments of their bakers. Is cake baking therefore a religiously neutral activity? As far as the motives and attitudes of the bakers are concerned, certainly not; but in its technical aspects, yes. Both the technical procedures and the material products are necessarily independent of the religious commitment of the baker. We gain nothing by attempting to claim otherwise.

So also with natural science when it is practiced and reported within the limits of its domain. While our religious commitment should directly affect our motivations, attitudes, integrity, and purpose for carrying out a scientific investigation, the technical aspects of scientific work are relatively impervious to religious bias. When an astronomer, for example, records the wavelength spectrum of starlight or interprets that spectrum in terms of thermal radiation and atomic absorption lines, his or her religious orientation has no control over the results. When astronomers measure the distance to a star, compute its lookback time, and make a conclusion concerning cosmic chronology, the results ought not to be biased by their religious perspective on the status, origin, or governance of the cosmos. The natural sciences, when functioning within the boundaries of their domain of legitimate operation, are religiously blind; they can neither confirm nor invalidate the concept of divine action.

Scientific theories pertaining to the internal affairs of the material world, like cakes baked according to a recipe, have been isolated from those aspects of human activity that reveal the religious commitment of the scientist. Consequently, we cannot use the claim of religious bias as a basis for either ignoring or discrediting the results of contemporary natural science—especially not the physical sciences such as physics, chemistry, geology, and astronomy. On the contrary, as Christians we should be seeking to employ all that natural science has learned about the Creation so that we might be better stewards in faithful service to the Creator. As stewards of God's Creation, some of us are professionally engaged in a disciplined investigation of its properties, behavior, and history. Our goal in this enterprise ought not to be the development of an *alternative* natural science, much less an *alternative to* the natural sciences; rather, we should set ourselves to the important and rewarding task of performing our scientific work in a *stewardly* fashion and with the highest standards of professional competence and integrity.

CHAPTER ELEVEN

More Heat Than Light: The Creation/Evolution Debate

The creation/evolution debate is alive and vigorous. The question before us, we are told, is "How did we get here?" Are we, along with the rest of the universe, the products of creation or evolution? Legislators, judges, teachers, preachers, school board members, and other community leaders are being asked to make policy decisions related to the issues raised by this question. Courtrooms, classrooms, school auditoriums, and church basements have become the sites of heated debate on the creation/evolution question. Before beginning an evaluation of the debate or of the positions held by the two principal opponents, perhaps we ought to ask a very elementary question: Why is there a debate at all?

The answer, many say, is really quite obvious: creation and evolution are inherently contradictory concepts. They are contradictory answers to the crucial question of origins, and they cannot be reconciled. Therefore, adherents to one view are necessarily opposed to the other, and debate is inevitable. But I would like to suggest that the debate is in fact *not* inevitable. To get at the heart of the issue, let's sharpen the focus of the question and examine the basis on which creation and evolution are perceived as contradictory and irreconcilable concepts.

In the first place, many people (perhaps a majority) assume that the term *creation* denotes the instantaneous inception by divine fiat of galaxies, stars, planets, and living creatures in fully developed and functioning forms, essentially the same forms that we see today. Second, many people (perhaps most) assume that the term *evolution* denotes a naturalistic process by which all bodies, both celestial and terrestrial, both animate and inanimate, have spontaneously and autonomously developed from unformed matter or from earlier forms without divine action or direction. Thus the contemporary debate usually boils down to a contest between *special creation* (instantaneous inception of a mature, functioning universe by divine fiat) and *naturalistic evolution* (continuous development of forms by autonomous material processes). However, unless creation does necessarily entail the concept of the instantaneous inception of mature forms and evolution does necessarily entail the concept of autonomous, naturalistic processes, the contemporary debate is being conducted in vain.

HOW TO LISTEN TO THE DEBATE

For most people, I suspect, it is difficult to listen to the creation/evolution debate in a disinterested manner. Many listeners are too strongly biased toward one position to entertain the other; many others, myself included, judge that neither position gets to the heart of the matter. As a Christian who wants to take both the Bible and the Creation seriously, I strongly desire to see biblical teachings accurately portrayed and the results of scientific investigation appropriately applied. The creation/evolution debate as it is most commonly conducted does neither. In a typical debate the concept of creation being presented has been stripped of most of its theological significance, so that what is left no longer portrays the biblical teaching with accuracy. And in most cases the concept of evolution being presented has been extrapolated so far beyond the boundaries of natural science that it is no longer an appropriate application of scientific investigation. The word *creation* has been given a diminished meaning while the word *evolution* has been assigned an inflated meaning. These are, I realize, harsh judgments; I will try to provide warrant for them later in our discussion.

First, however, we should clarify a number of important distinctions. The faults that I find in the debate are in large part a direct consequence of failures to define and use words carefully. Careful distinctions between similar words and precise delineations of diverse meanings assigned to the same word are of utmost importance. Therefore, in preparation for listening to the creation/evolution debate with discernment, let us briefly discuss a number of crucial distinctions.

Categories and Sources

Investigation of the creation/evolution debate soon reveals that it spans a broad spectrum of issues—some philosophical, some theological, some scientific. Within the realm of science, the debate deals with questions that are drawn from the domains of cosmology, astronomy, physics, chemistry, geology, biology, anthropology, and a host of related disciplines. Each side of the debate asks a considerable range of questions and provides its own answers.

The quality of the entire debate and the credibility of the argumentation on either side depend critically on the care with which the questions are properly categorized and addressed to appropriate sources. To judge the validity of the arguments, we have to identify each question individually, determine whether it is being directed to an appropriate source, and then evaluate the answer in terms of that source. In the case of questions about the material world, for instance, we should note carefully the category into which it falls (the eight categories and the internal/external classification we considered in Chapter Ten should prove very helpful in this process) and then ask whether a question in that category can appropriately be addressed to that source. In short, I am recommending that the creation/evolution debate be judged by applying the principles of categorical complementarity. Not only should each position be tested according to these principles, but the structure of the whole debate should be judged in light of the careful distinctions demanded by that approach.

Either/or-manship: The Fallacy of Many Questions

As I have suggested, the contemporary creation/evolution debate has on the whole come down to a contest between special creationism and naturalistic evolutionism. There are other positions than these, of course, but none of them has achieved the sort of popular acceptance that the two principal positions have—and indeed, not a few proponents of those two principal positions specifically deny the validity of any other points of view. Henry Morris, for example, has said,

> The fact is . . . that there are *only two* possible models of origins, evolution or creation. . . . Either the space/mass/time universe is eternal, self-existent, and self-contained, or it is not. If it is, then evolution is the true

> explanation of its various components. If it is not, then it must have been created by a Creator.[1]

And Duane Gish makes essentially the same point when he concludes that there are

> two views of origins that are logically and philosophically consistent: either our universe and the living things it contains were created and their origin was miraculous, or they arose mechanistically from disordered primordial stuff by a process of self-transformation. . . . Either man was created, or he evolved.[2]

There is, however, a serious logical fallacy in this either/or format, commonly called the "fallacy of many questions." Both of the principal positions in the debate are made up of answers to several diverse and categorically distinct questions, each of which should be individually investigated. By the either/or format of the debate, however, we are prevented from making such investigations. We are simply offered two packages of answers and the demand that we choose one or the other. This is a crucial error, because, as we will see, it commits us to mixing up specific answers to questions of internal affairs (cosmic properties, behavior, and history) with specific answers to questions on matters of external relationship (cosmic status, origin, governance, value, and purpose). To demand a choice between only two packages of answers to distinct and separable questions is to commit the fallacy of many questions.

The essentially misbegotten format of the creation/evolution debate presents us with a false dilemma. Many Christians—myself included—find neither of the two choices acceptable, and indeed object to the very format of the debate. Furthermore, the debate is leading many non-Christians to identify the Christian faith with a perspective on material behavior and cosmic history that I judge to be deficient both theologically and scientifically. That disturbs me greatly. Everyone, both within and outside of the Christian community, deserves to know that there is a biblically based perspective on the cosmos that avoids the pitfalls of both naturalistic evolutionism and theistic antievolutionism (special creationism).

---

1. In Henry M. Morris and Gary E. Parker's *What Is Creation Science?* (San Diego: Creation-Life Publishers, 1982), p. 156.

2. Gish, "It Is Either 'In the Beginning, God'—or . . . Hydrogen,'" *Christianity Today*, 8 October 1982, p. 29.

Origins, Origins, and Origins

After the words *creation* and *evolution,* one of the most frequently used terms in the debate is *origin.* I have already used this term in formulating the list of categories of questions that can be asked concerning the material world. Like the other key terms in the debate, *origin* is used in different contexts with substantially different meanings, and we would do well to make careful distinctions among them at the outset of this discussion.

The term *origin* is frequently used to designate the beginning of something's existence, to designate, in other words, a thing's *temporal origin.* At other times it is used to denote the manner in which something was formed from existing materials—to designate its *formal origin;* in such contexts the word *formation* would be a suitable, perhaps preferable, synonym. The term is also used in a fundamental and profound sense to designate the source or cause for the existence of something—not merely its beginning, not merely its formation, but the very source of its continuing existence, the "ground" of its being. It was this sense of *ontic origin* that I referred to in one of the categories of questions concerning the relationship of the Creation to the Creator, when I suggested that the Creator continuously acts as the origin, or cause, or source, of the world's continuing existence.

There are other meanings, but these three—temporal, formal, and ontic—are the most important in this discussion. They point to different categories in the classification scheme we have adopted. Questions of temporal and formal origins ordinarily belong to the internal categories of cosmic history and behavior and so would call for investigation by the methods of natural science. Questions of ultimate causation or ontic origin, on the other hand, belong in the category of external relationships and so should be addressed to the source of one's worldview—either Scripture or metaphysical speculation. We will go a long way toward clarifying the creation/evolution debate if we can remain clear on whether the term *origin* is being used in its temporal, formal, ontic, or yet another sense in any given argument.

Lawful Behavior and Governance

We have talked at length about the ordinary patterned behavior of matter and material systems. Atoms, for example, behave and interact according to certain orderly patterns. Atomic behavior in turn imposes patterns on the behavior of larger material systems, such as stars. And stellar behavior brings about changes in stellar properties—the sequence of changes that we call "stellar evolution."

Material behavior can be investigated empirically—that is, we can freely observe what matter and material systems do. Because the observed behavior of matter follows very strict patterns which can be precisely described and mathematically represented, we often refer to these patterns, or their mathematical representations, as "natural laws." Matter, we say, behaves in a lawful (i.e., patterned) manner.

But what do we mean by the term *natural law*? If we mean simply a descriptive statement about the patterns of behavior exhibited by matter and material systems (natural behavior), well and good. That would be an acceptable and convenient term to identify that concept. In fact, let us agree that from this point on we will use the terms *natural law* and *natural behavior* in that empirical and descriptive sense. Natural science, then, can be identified with the investigation of natural behavior and the natural laws that describe it.

We commit a serious error, however, if we change the status of natural law from a description of material behavior to the governor of that behavior. In virtually every science textbook, and in most literature concerning the phenomena within the domain of natural science, one reads not only about the natural laws *describing* material behavior but also about the laws *governing* nature. Unfortunately, one can readily find examples of this language both in naturalist literature—

> Hand in hand with our progress in observing the universe in more detail and in new ways must come increasingly refined theories, capable of organizing the flood of data into a comprehensive framework of understanding. This framework is based upon the laws that describe how matter behaves. . . . The gods controlled the universe of the ancients; the laws of physics control ours. . . .In order to understand the universe, we must understand the laws that govern it.[3]

—and in creationist literature—

---

3. Robert V. Wagoner and Donald W. Goldsmith, *Cosmic Horizons* (San Francisco: W. H. Freeman, 1982), pp. 32-33.

> In this chapter we wish particularly to look at the two models [creation model and evolution model] and their implications concerning the basic laws of the physical universe, especially those which govern the nature of the possible changes which can occur in the systems and processes of the universe.[4]

What on earth does it mean to say that a law *governs* material behavior? A law, whether civil or natural, is not a governing agent; it is no more than a statement of the pattern of behavior that the governing power imposes. A natural law is simply a statement that describes the pattern of behavior enforced by the governor. It is important that we distinguish lawful behavior from lawful governance in any discussion, and especially in the creation/evolution debate. As we have already noted, behavior is a category classified under internal affairs, and governance involves the consideration of external relationships. Behavior is just a matter of what happens; governance is a matter of the ultimate causation of what happens. Behavior is a topic in physics; governance is a question for metaphysics or theology.

Natural and Naturalistic

If "natural behavior" is simply the way things ordinarily occur, and "natural laws" are just statements describing natural behavior, what is meant by the term "naturalistic behavior"? Unfortunately, the term is not used in an entirely consistent manner, but most commonly, and certainly in the present discussion, it is meant to incorporate both natural behavior *and* the specification that the governing agent is matter itself. Thus, "naturalistic behavior" is *autonomous* natural behavior—matter behaving according to self-caused and self-governed patterns.

It is the specification of autonomous self-governance that makes naturalistic behavior such a different concept from natural behavior, which entails no identification of the governing agent. Essentially the same distinction is at work here as the one that we made in distinguishing between natural science and naturalistic science. In the former, the question of governing agent is purposely placed outside of its scope; in the latter, the governing agent is identified as matter itself, and the existence of

---

4. Morris and Parker, *What Is Creation Science,* p. 158.

any external agent is explicitly denied. By contrast, "creationomic science" identifies God as the governing agent in all material behavior.

In listening to the creation/evolution debate, we must be alert to the natural/naturalistic distinction to be certain that it is clearly and consistently employed. Whenever the distinction between these two terms is either carelessly or purposely omitted, the crucial question of governance cannot be treated adequately, and the debate soon becomes meaningless, a waste of both time and effort.

Natural Processes and Divine Action

Failure to distinguish between *natural* and *naturalistic* leaves one highly vulnerable to another either/or fallacy. If we don't distinguish natural processes (the ordinary patterned behavior of matter that we observe) from naturalistic processes (natural processes perceived as self-governed without the need for divine governance), we can easily slip into the trap of placing natural processes and divine action in competition with one another. The creation/evolution debate is permeated with such confusions. Both sides characteristically insist that something must have happened either as a result of natural processes or as a result of divine action:

> The origin and development of all things can either be explained in terms of natural processes which are continuing to operate today—or they cannot. If not, then non-natural processes (or extra-natural, preternatural, or supra-natural, if you prefer) must have operated in the past to originate and develop at least some of the contents of the universe, as well as the universe itself.[5]

This assumption that natural processes somehow compete with or constitute an alternative to divine action betrays an unbiblical concept of natural process. Biblical theism stresses that the Creator is not only the Originator of Creation but also its Preserver, Governor, and Provider. There is, in other words, no natural process that falls outside of the Creator's domain of action. What we call a natural process is not something that stands outside of his control; it is, rather, a display of his governance, a manifestation of his sovereignty. Whatever the Creator origin-

5. Morris and Parker, *What Is Creation Science*, p. 156.

ates, he maintains in service to his purposes. To hold that natural processes are beyond divine direction is not theism but deism. To hold that natural processes compete with divine action is dualism. To hold that natural processes are a substitute for divine action is naturalism. Anyone who wants to take the Scriptures seriously will reject the misconceptions of natural processes drawn from naturalism, dualism, and deism.

By embracing biblical theism we see all phenomena occurring under divine direction and serving divine purposes. Stellar evolution and the opening of an almond blossom are equally the display of divine governance. Quasar outbursts a billion years in the past and today's sunshine are equally a manifestation of God's sovereign rule over his Creation.

Natural and Supernatural

One common way to identify some phenomenon as occurring by divine action is to call it a "supernatural" act. But this label is most often used to describe phenomena that for one reason or another stand outside the realm of natural phenomena, events, or processes. Natural and supernatural, then, are often presented as being mutually exclusive categories. As John C. Whitcomb, Jr., explains,

> In forthright opposition to all efforts to explain the origin of the world in terms of purely natural processes, the Bible states that God created all things supernaturally. . . . It [the earth] was created by purely supernatural means during six literal days and completely furnished with all the basic kinds of living things that have ever existed, including man.[6]

Where references to supernatural acts are intended simply to remind the reader of the immanence of God in the events of this world or to point to the extraordinary character of some divine acts, I offer no criticism. But whenever it is suggested that there are two categories of phenomena in the cosmos—some natural, others supernatural—we must be very careful. If we draw the line on the assumption that creation requires divine action but natural processes do not, then we have slipped into the pit of deism; the unavoidable implication is that God is re-

---

6. John C. Whitcomb, Jr., *The Early Earth* (Grand Rapids: Baker, 1972), pp. 21, 135.

quired for initiating the cosmos, but once initiated it can run on its own. The deist contends that the universe needed a divine push to get it rolling, but having gotten started, it can now coast naturalistically.

Adjacent to the pit of deism is the quicksand of interventionism. According to that perspective, most things in the material world happen "naturally" (in essence, naturalistically), but on certain special occasions God breaks into this realm and supernaturally intervenes in the affairs of the material world or its creatures. Once again, I certainly do not wish to diminish or negate the reality of divine action in the cosmos; indeed, I want to stress that biblical theism requires that we be aware of his constant active presence. According to Scripture, God need not "intervene" or "break into" the natural machinery of the cosmos as if it were already running independently of him. God's active presence is required at all times, not just on special occasions that demand supernatural intervention.

Creation and Instantaneous Inception

The most vocal participants in the creation/evolution debate (including representatives of both sides) assume that the word *creation* means, almost exclusively, the sudden appearance of all manner of objects. For the special creationists, the concept of creation as instantaneous inception is central:

> An important aspect of the supernaturalism of the original creation was its *suddenness.* Creation was not only *ex nihilo* (in reference especially to the earth, the sun, the moon, and the stars), but it was also, in the very nature of the case, instantaneous.[7]

Or, as Morris and Parker have put it,

> At some point of time, say T_0, the Space/Mass/Time cosmos was simply *created,* brought into existence in fully developed and functioning form right at the beginning. The complex structures of its immense variety of stars and galaxies did not *evolve* at all. They were simply created, with any changes since that time limited to the processes of decay, not development.[8]

---

7. Whitcomb, *The Early Earth,* p. 24.
8. Morris and Parker, *What Is Creation Science,* p. 228.

From quite a different perspective, Robert Jastrow also identifies the word *creation* with instantaneous initiation, such as one associates with the big bang:

> A sound explanation may exist for the explosive birth of our universe; but if it does, science cannot find out what the explanation is. The scientist's pursuit of the past ends in the moment of creation. . . . For the scientist who has lived by his faith in the power of reason, the story ends like a bad dream. He has scaled the mountains of ignorance; he is about to conquer the highest peak; as he pulls himself over the final rock he is greeted by a band of theologians who have been sitting there for centuries.[9]

While these authors hold quite different concepts of the end product of creation (a fully developed cosmos functioning essentially as it is today versus a primeval fireball that evolved into the cosmos we now observe), they do hold in common the idea that "creation" means instantaneous initiation. In the popular vocabulary, creation is a singular act, or event; it happened once, and now it is completed.

Biblical theism, however, holds no such emaciated view of creation. To know God as Creator is not merely to know of one of his acts in the past. To know God as Creator is to experience one's continuing relationship to him—past, present, and future. As Calvin has expressed it,

> To make God a momentary Creator, who once for all finished his work, would be cold and barren, and we must differ from profane men especially in that we see the presence of divine power shining as much in the continuing state of the universe as in its inception. (*Inst.*, 1.16.1)

According to biblical theism, the word *creation* stands not merely for an instantaneous act or event but for an eternal, covenantal relationship. Even an event as important as an initial act of exnihilation—if that is a valid concept—stands as but one grain of sand on the endless seashore of the Creator's eternal relationship with his Creation.

As Christians, we should never accept a diminished view of creation, because from that will flow a diminished view of the Creator. Any view of creation that so emphasizes or isolates an act of inception as to minimize the attention paid to the

---

9. Jastrow, *God and the Astronomers* (New York: Warner Books, 1978), pp. 104-6.

continuing relationship of the Creator to his Creation is a view that bears more resemblance to the theologically bankrupt perspective of deism than to the abundantly rich viewpoint of biblical theism. If that band of theologians greeting Jastrow's intrepid scientist speak of the Creator as no more than an Originator and of creation as no more than an act of instantaneous inception, then they are not biblical theists but merely deists. The God of deism is a distant God whose acts are confined to the remote past. Deists remember God's transcendent acts but fail to experience his immanent activity. Instantaneous inception is no fitting substitute for an eternal covenant relationship.

Furthermore, for a person committed to a naturalistic worldview, instantaneous inception constitutes no proof of divine activity. If it is called for by empirical or theoretical considerations, the instantaneous appearance of matter or energy or spacetime will simply be one more item on the list of naturalistic phenomena. Fred Hoyle's steady-state cosmology, which has received serious consideration in past decades, postulates the continual creation of hydrogen atoms. The phenomenon of instantaneous inception has no theological significance in itself; it is just another naturalistic process as far as the proponents of steady-state cosmology are concerned. More modern evolutionary cosmologies describe a universe that has developed to its present state from a singular inception—the big bang, which appears to have occurred about fifteen billion years ago. While some people are tempted to identify the big bang as "the creation," others see no need for any reference to divine action at all. According to some naturalists, our universe may be no more than a large fluctuation in the "vacuum state," a fascinating naturalistic phenomenon, but no more. As before, instantaneous inception is an empty substitute for a dynamic, eternal, covenant relationship between the Creator and his Creation.

THE SHORTCOMINGS OF NATURALISTIC EVOLUTIONISM

Its Naturalistic Worldview

It is essential to understand that the opponent of special creationism in the contemporary creation/evolution debate does not merely argue for the evolutionary character of cosmic history but rather for a strain of naturalism that has elevated the material processes of evolution far beyond the categories of behavior and history. Regrettably, much of the contemporary debate is carried on in the arena of natural science, as if public debates before general audiences will settle highly technical questions about the interpretation of empirical data or the evaluation of theoretical models. In fact, however, the questions lie not within the categories of internal affairs but in the categories of external relationships—status and its consequences for origin, governance, value, and purpose. Thus, in the remainder of our discussion we will limit ourselves primarily to these issues.

The foundation of modern Western naturalism is its presupposition that the ultimate reality is matter and the spacetime arena in which it operates: there is nothing beyond the material world, no category of divine being, only self-existent matter. Or, to express it differently, according to naturalism, the cosmos has the status of the one and only reality; it stands alone—not as a deity, as in ancient paganism, but as a machine-like substitute for deity. Thus, naturalism is essentially reductionistic, insisting that material systems are nothing but material systems; man is nothing but a marvelous molecular machine.

Ancient Eastern paganism often assigned the cosmos the status of deity; modern Western naturalism assigns the cosmos the status of Nature. Nature (with a capital *N*) is presumed to be the ultimate, self-existent, and sole reality. From this concept of the status of the cosmos, several consequences follow.

1. As Nature, the cosmos has its origin in itself; Nature exists independently of any external, nonmaterial being for the simple reason that there is no other being outside of Nature. No other power or person is presumed necessary for either the beginning or the preservation of Nature's existence. Nature is held to be self-originating and self-sustaining.

2. As Nature, the cosmos is autonomous, self-governed. Everything that happens in Nature happens solely as a consequence of the self-directed behavior of matter. No other power or person is required for the governance of autonomous Nature.

3. As Nature, the cosmos derives no value from its external relationships; in point of fact, there are no such relationships. If the cosmos or anything within it is to have value, it is held that it will have to derive that value from within the material world itself. There exists no source of value external to Nature.

4. As Nature, the cosmos functions without purpose. There is no goal toward which it is being directed. The superficial appearance of purpose in isolated instances is entirely accidental and is not part of a comprehensive plan. Though locally or temporarily there may be orderly patterned behavior and pro-

gressive temporal development, that must be seen as a transient effect—random noise that might momentarily be mistaken for a meaningful voice. Discussion of purpose must be limited to these transient phenomena.

Its Unacceptable Reasoning

The perspective of naturalistic evolutionism develops when one interprets the evidence of the natural processes of evolution (spatial, elemental, galactic, stellar, planetary, biological, etc.) within the framework of the worldview of naturalism. Committed to the naturalistic worldview, its adherents do not recognize biblical revelation as a legitimate source for answers to questions of status and its consequences. Such questions are addressed, rather, to the human mind and its metaphysical speculations. The material processes of evolution are elevated from a descriptive to an explanatory status, and evolution*ism* is the consequence.

It is my judgment that naturalistic evolutionism generally fails to distinguish carefully among the diverse categories of questions about the material world and consequently fails to address important questions to the proper sources for answers. The fundamental question of status is directed to metaphysical speculation for an answer. The question of ontic origin is either ignored or declared unanswerable. The question of governance is usually confused with the question of behavior. Questions of value and purpose are directed to the material world itself.

Those who fail to distinguish questions of behavior from questions of governance are particularly vulnerable to the error of claiming that evolutionary processes are necessarily naturalistic processes. And of course this is the central theme of naturalistic evolutionism—that the natural processes of evolution may be perceived as a substitute for divine action. All material processes, evolution included, are perceived as autonomous.

The biblical theist, however, is well prepared to avoid this trap. Knowing that all material phenomena are divinely governed, the theist recognizes that there is no logical relationship whatsoever between that pattern of material behavior that we call "evolution" and the claim that such a process is inherently naturalistic. The theist knows that *no* natural process is inherently naturalistic.

As but one example of the patterned behavior of material systems, evolution deserves no special status. If evolutionary processes are necessarily naturalistic, our everyday weather must be naturalistic as well. If evolution is inherently naturalistic, the processes of physical development that we observe when a child grows to maturity must be naturalistic too. If, on the other hand, we are able to perceive a winter snowfall as a manifestation of divine governance, we should also be able to perceive stellar evolution as divinely directed. And if we are able to perceive the birth and growth of a child as evidence of God's faithfulness and providence, we should also be able to perceive planetary evolution as evidence of God's purposeful guidance of his Creation.

Quite simply, the claim that evolutionary processes are inherently naturalistic is unwarranted. No material process is *necessarily* naturalistic. The claim may be consistent with a naturalistic or deistic conception of the cosmos as Nature, but it is entirely inconsistent with the biblical concept of the cosmos as Creation: there is no material process in Creation that falls outside of divine governance. Consequently, when naturalistic evolutionists and theistic antievolutionists agree to label evolution as an inherently naturalistic process, they are both in error. And if evolutionary processes are not inherently naturalistic, then it should be clear that it is an error to assume that naturalistic evolutionism is the only alternative to special creationism. Neither theists nor naturalists should be willing to accept such an obvious failure to separate a complex issue into its categorically distinct parts.

Its Theologically Barren View of Creation

While naturalistic evolutionism has an inflated concept of evolution, it generally expresses a deflated concept of creation. Within the context of the creation/evolution debate, the term *creation* is used primarily to denote "instantaneous inception," with or without an apparent cause. However, as we have already pointed out, such a concept of creation is theologically barren, bearing little resemblance to the biblical concept of creation as a dynamic continuing *relationship* rather than a single, isolated, discontinuous *event.*

Naturalism tends to have a deistic concept of divine action in the cosmos. With its emphasis on creation as an initiating event, naturalism often fails to consider the implications of preservation, governance, and providence. By limiting its concept of creation to the temporal matter of beginnings, naturalistic evolutionism overlooks the need to deal with the question of ontic origin, which requires an explanation for both beginning *and* continuation. And of course naturalistic evolutionism rejects the possibility of any sort of divine action beyond the point of the unknowable initiating event.

As Christians, we must be particularly alert that biblical concepts and terms be accurately represented and properly used. When naturalism uses the term *creation* in an unbiblical and theologically bankrupt manner, we must challenge that misuse of our term. Even more importantly, we must refuse to adopt the deflated concept of creation as instantaneous inception and insist on using the full biblical concept of creation as a continuing dynamic relationship. No productive debate can be carried out until the participants come to a mutual understanding of the terms and concepts that lie within the scope of the discussion.

THE PITFALLS OF SPECIAL CREATIONISM

Indentifying the Position

As we have noted, the chief alternative to naturalistic evolutionism in the contemporary creation/evolution debate is special creationism—a view that not only specifies the relationship of the cosmos to its Creator but also prescribes a detailed picture of the manner and chronology of the divine action that brought the cosmos into being. An integral component of special creationism is a dependence on Genesis 1 as a source of answers to questions of cosmic history. Creation is seen as a series of discrete divine acts performed during six twenty-four-hour periods a few thousand years ago. The end product of these acts of "special creation" is the fully developed and functioning universe that we see today—a cosmos populated with planets, stars, galaxies, and quasars distributed over billions of light-years of space. Though these objects may have the appearance of great age, special creationists maintain that they were instantaneously endowed with that appearance at their sudden inception by divine fiat. In addition, most special creationists hold that the divine acts that gave the cosmos its present form and structures were entirely different in character from the divine action that now preserves and governs the universe.

The desire of special creationists (or theistic antievolutionists, if you prefer) to have their viewpoints taught as the "creation model" in the public school science curriculum has led to the development of two versions of special creationism—"biblical special creationism" and "scientific special creationism." The concepts of the chronology of cosmic history and the character of divine creative activity are essentially the same in these two versions, but they seek to warrant these concepts by quite different means. Biblical special creationism seeks to warrant its concept of creation by biblical exegesis. Scientific special creationism seeks to warrant the concept of creation by means of empirical investigation, the type of study that can be taught in the public school system. In the discussion that follows, I'll use the term "special creationism" to refer to both versions and treat them separately only where a distinction is significant.

Before directing a series of criticisms to the position of special creationism, I must say at the outset that there are numerous important points on which I find myself in agreement with special creationists. We can agree, of course, that God is the sovereign Creator and that the whole cosmos is his Creation. From this fundamental agreement concerning the status of the material world there follows agreement on many (though not all) of the consequences of that status in matters of origin, governance, value, and purpose. Furthermore, special creationists and I are in total agreement on the unacceptability of a naturalistic perspective on the status of the cosmos. Together we believe that the cosmos is not merely Nature but Creation.

Where, then, do differences arise? Principally from the assertion by the proponents of special creationism that evolutionary processes are inherently naturalistic and that creation necessarily entails the instantaneous inception of objects in fully developed and functioning forms. It is the combination of these two positions that forms the foundation for the either/or format of the creation/evolution debate to which I have strong objections.

Its Naturalistic Concept of Evolutionary Processes

When reading the literature of naturalistic evolutionism, one expects to find the claim that evolutionary processes, like all other material processes, are inherently naturalistic—not divinely directed, but self-governed. It would come as no surprise, therefore, to find a statement such as the following in a book written by someone who describes himself or herself as a "reductionist-evolutionist."

> The outstanding mystery now seems to me to be the origin of matter—why matter exists (has evolved?) capable of forming itself into the most complex patterns of life. It is a fundamental evolutionary generalization that no external agent imposes life on matter. Matter takes the forms it does because it has the inherent capacity to do so.[10]

What does surprise me, however, is to find Christians who are willing to grant that any material process could possibly be naturalistic, or autonomous. Yet the special creationist literature is shot through with claims that evolutionary processes are inherently naturalistic. Duane Gish, for example, makes the claim in no uncertain terms:

> Since evolutionary theory is an attempt to explain origins by a process of self-transformation involving only naturalistic and mechanistic processes, God is unnecessary and so excluded from the process. . . . By definition, evolution is a strictly mechanistic, naturalistic, and, therefore, atheistic process.[11]

In the context of this assertion that evolutionary processes necessarily exclude divine action (though other natural processes presumably do not) the either/or format of the debate is inevitable. Furthermore, once special creation and naturalistic evolution are set up as the sole and mutually exclusive alternatives, evidence against one automatically becomes evidence favoring the other. As Henry Morris explicitly puts it, "Since there are only two possible models, and they are diametrically opposed, it is clear that evidence against evolution constitutes positive evidence for creation and evidence against creation is evidence for evolution."[12] This may explain why so much of the energy of the special creationist organizations is devoted to the production and dissemination of antievolutionist literature: any argument against any aspect of evolutionary theory or evolutionary processes is perceived as an argument for the reality of divine action as an alternative to, or substitute for, natural processes. If evolutionary processes are inherently naturalistic, then

---

10. P. J. Darlington, Jr., *Evolution for Naturalists* (New York: John Wiley, 1980), pp. 232-33. The term *naturalist* in this title means simply a careful observer of nature. I am not inferring Darlington's philosophical naturalism from the title but from a number of forthright statements of perspective he makes.

11. Gish, "It Is Either . . . ," pp. 28-29.

12. Morris, in *What Is Creation Science,* p. 162.

any God-fearing believer must be opposed to such an atheistic concept.

But are evolutionary processes necessarily naturalistic? Is there any warrant for excluding this one category of natural processes from the domain of divine governance?

In a number of different ways, we have already suggested that evolutionary processes are not distinguishable from other material processes. Evolutionary processes are those ordinary natural processes that display the ordinary patterns for material behavior and lead to the temporal development of any material system. Stellar evolution, for example, follows directly from the effects of gravity, thermodynamics, energy transformation, nuclear interactions, electromagnetic radiation, and a host of other ordinary patterns of material behavior. To speak of stellar evolution requires no appeal to extraordinary or unusual or unnatural phenomenon. To claim that stellar evolution is inherently naturalistic is equivalent to claiming that all natural processes are necessarily naturalistic. And we know that no natural process ought ever to be viewed as inherently naturalistic.

The biblical theist knows that to affirm the authenticity of the regular patterns of material behavior that we can empirically discover in the Creation does not entail the denial of the reality of divine governance. The biblical theist perceives, through the spectacles of Scripture, that these orderly patterns of behavior are in fact a revelation of the faithful and lawful governance of the material world by the Creator. The patterned behavior of matter must not be viewed as a substitute for or an alternative to divine action; it must be perceived as evidence for divine governance. The Creator revealed by Scripture is at work at all times and in all places within his Creation—even in distant stars and galaxies, even in the remote past that our minds can scarcely comprehend.

Looking at this behavior/governance relationship from the other side, the biblical theist also knows that to affirm the reality of divine action in the cosmos does not entail the denial of the regular patterns of material behavior exhibited in natural processes. Accepting the immanence of divine governance does not require, or even suggest, the denial of the authenticity of patterns in natural behavior or the validity of proximate cause-effect relationships.

To affirm, for example, that God is my Maker, that he "knit me together in my mother's womb" (Ps. 139:13) does not require, or even suggest to me, that I should deny the possibility that there is in effect a patterned behavior of matter through which a fertilized egg develops into a mature human being.

And to affirm that God is the great Healer who "heals all of our diseases" (Ps. 103:3) does not require, or even suggest to me, that I should deny the possibility that there are patterns in material behavior that will make medical and surgical treatments effective. Our newspapers frequently inform us of the tragic consequences that follow parents' decisions to withhold medical treatment from children because they "trust that the Lord will heal." To deny available medical treatment is to reject the means that the Lord is providing. And to think of divine activity only in terms of "supernatural intervention" is to close one's eyes to the Lord's unceasing work throughout the whole universe—in what we call "natural processes." Is it possible that the demand for spectacular forms of divine intervention actually betrays a lack of belief in divine governance and a lack of faith that God can accomplish his purposes through "natural" processes?

To affirm that God is the Lord of clouds, lightning, rain, and wind (see Ps. 135:6-7) does not demand, or even suggest to me, that I should deny the possibility that the patterned behavior of matter serves as the proximate cause for all of these meteorological phenomena. To affirm the reality of God's governance of atmospheric phenomena does not make the science of meteorology unnecessary or un-Christian. The Christian meteorologist has just as much need of the atmospheric sciences as does the naturalistic meteorologist. We have no need for specifically "Christian meteorology," using biblical barometers and theistic thermometers and angelic anemometers and regenerate radar instruments; we simply need good, honest, religiously neutral meteorology.

Finally, to affirm that God is the Creator of the stars, who has made them "by the breath of his mouth" (Ps. 33:6) does not demand, or even suggest to me, that I should deny the possibility that there are orderly patterns of material behavior that will function in such a way as to lead globules of interstellar gas to gravitationally collapse into main-sequence stars. Nor is there any reason to deny that thermonuclear transformations in stellar cores will cause main-sequence stars to evolve into red giants, white dwarfs, neutron stars, or black holes. To affirm that God is the Creator of the cosmos does not demand the denial of stellar evolution. The temporal development of stars is just as much a consequence of divine action as is a winter snowfall or the medicinal healing of a disease or the birth of a child. Each of these phenomena appears to be a natural process when observed from the perspective of natural science, which deals only with what can be empirically detected and theoretically modeled. Yet at the same time, each of these phenomena can be seen as the manifestation of faithful divine governance when viewed through the spectacles of Scripture. Neither view demands, or even suggests, the denial of the other perspective.

Are evolutionary processes inherently naturalistic? No more so than snowfall, or healing, or childbirth. The claim made by naturalistic evolutionists and special creationists alike that evolutionary processes are inherently atheistic must be completely rejected by all who take both the Bible and the Creation seriously.

Its Denial of Authentic Cosmic History

With relatively few exceptions, the proponents of special creationism also defend the view that the cosmos is very young—with an age in the neighborhood of ten thousand years rather than many billions of years. There are some solid pragmatic reasons for coupling the young universe hypothesis with special creationism. All of the evolutionary scenarios that we discussed earlier make sense only in the context of the empirically derived time scale that places the duration of cosmic history at approximately fifteen billion years. Therefore, in the context of an either/or debate format, it would appear quite expedient to seek support for the "creation model" by attempting to discredit the standard time scale for cosmic history.

Some special creationists seek to find empirical warrant for their belief that the universe is actually no more than a few thousand years in age. Much of what has come to be called "scientific creationism," or "creation science," is dedicated to the goal of uncovering weaknesses in the standard methods of determining the age of things, and of substituting alternative procedures for age determinations—procedures that yield values closer to ten thousand years.

Other special creationists readily admit that the standard chronometric methods are reasonable and that they do indeed yield cosmic age values in the multibillion-year territory, but then they insist that this is merely an "apparent age," not the actual age. The defenders of the apparent-age hypothesis argue that if God did create a fully developed and functioning universe, then it would necessarily have the superficial appearance of age—it would necessarily look as if it were older than it actually is. Whitcomb makes this assertion very directly:

> The supernaturalism and suddenness of creation provide a necessary background for the concept of creation with a superficial appearance of history or age. . . . If this doctrine were not true, there could have been no original creation by God at all.[13]

And Henry Morris makes essentially the same assertion:

> Actually, real creation necessarily involves creation of "apparent age." Whatever is truly created—that is, called instantly into existence out of nothing—must certainly look as though it had been there prior to its creation. Thus it has an appearance of age.[14]

According to this special creationist argument, not only was God able to create a world with the appearance of antiquity and history, but he was also required to do so. Divine creativity of any other sort appears unacceptable to these defenders of the apparent-age concept; creation without apparent age is not "real" or "true" creation according to the proponents of the young universe theory.

In my judgment, the situation is quite the contrary. I find the apparent-age concept to be entirely unacceptable because it requires God's Creation to be more of an illusion than a reality. The crux of the matter is the distinction between age and history. Is the created cosmos that we observe merely endowed with the superficial appearance of age, or is it rather permeated with an authentic record of history? Is the inference of cosmic antiquity merely an unwarranted extrapolation from superficial appearances, or is it a legitimate conclusion drawn from an examination of the physical record of actual events? I am convinced that it is the latter. I am convinced that the age of the universe is no mere superficial appearance, and that cosmic history is no mere illusion.

You will recall that when we look at a distant celestial luminary we do not see it as it is at the present moment; rather, we are seeing light that left the body some time ago—many billions of years ago in the case of the objects most distant from

---

13. Whitcomb, *The Early Earth*, p. 29.

14. Morris, *The Remarkable Birth of Planet Earth* (Minneapolis: Dimension Books, 1972), p. 62.

us. The time delay, or lookback time, is equal to the amount of time required for light to travel from the source to the observer. For the sun, the lookback time is a mere eight minutes: when we look at the sun, we see the events that were occurring at the solar surface eight minutes earlier. When we observe the Andromeda galaxy we are observing events that occurred in that galaxy two million years ago. And when we look at the quasar 3C48, which is about five billion light-years distant, we are seeing events that took place about five billion years ago.

The question before us is whether these events are part of an authentic cosmic history or are no more than a divinely fabricated illusion. Interpreting these phenomena as mere "superficial appearances" makes no sense whatsoever. That possibility is clearly eliminated by the consequences of lookback time. The events, recorded in the light itself, must be a part of authentic history or else they never occurred at all.

The solution commonly offered by young-universe special creationists is that God created not only light sources but also a full set of electromagnetic waves reaching from the sources to the earth, appearing to the terrestrial observer as if they had been produced by specific luminous sources. In *Scientific Creationism,* Henry Morris proposes such a scenario:

> This fact means that the light from the sun, moon and stars was shining upon the earth as soon as they were created. . . . As a matter of fact, it is possible that these light-waves traversing space from the heavenly bodies to the earth were energized even *before* the heavenly bodies themselves in order to provide the light for the first three days. It was certainly no more difficult for God to form the light-waves than the "light-bearers" which would be established to serve as future generators of those waves.[15]

But consider the implications of such a proposal. If the cosmos is only a few thousand years old, and if the light we are now receiving was created en route to appear as if it were coming from a distant source, then most of the visible universe and nearly all of cosmic history is reduced from reality to illusion.

The light that appears to come from sources more than a few thousand light-years from us would bear no causal relationship to any actual luminous objects. The apparent sources

15. Morris, *Scientific Creationism* (San Diego: Creation-Life Publishers, 1974), p. 210.

need not even exist. All but a small part of our Milky Way galaxy would be no more than an electromagnetic illusion—an intricately detailed deception—complete with information about chemical compositions, surface temperatures, motions, luminosity values, and a host of other details about things whose existence is entirely irrelevant. And all of the light-producing events that appear as if they had happened millions or billions of years ago would not have actually occurred. All but the last few thousand years of a multibillion-year cosmic history would be no more than an illusion. If the apparent-age hypothesis were true, then that magnificent, orderly, coherent history of which we spoke earlier would necessarily be reduced to a divinely perpetrated hoax. The apparent-age hypothesis, an integral part of special creationism, substitutes deceptive illusion for authentic, divinely directed cosmic history.

To consider just one more example of how the apparent-age hypothesis transforms authentic cosmic history into illusion, recall our discussion of the properties of stellar clusters. The stars of a cluster are not merely stars that are found in close proximity to one another; they are also distributed on an H-R diagram in a very peculiar way—displaying a pattern that clearly reveals a particular history that has been experienced by that cluster. The question is whether that history is authentic or illusory. Special creationist Paul Steidl recognizes that stellar clusters pose a significant problem for proponents of the young-universe theory, but he offers no solution.

> Perhaps the most important remaining question for [special] creationists is the origin of the turnoff points in the H-R diagrams of different clusters. The stars are real physical objects and presumably follow physical laws; we would rather not take the easy way out by saying simply "God made them that way." But if [special] creationists take the position of rejecting stellar evolution, they should provide a feasible alternative.[16]

For young-universe special creationists, however, there is no alternative except to propose the apparent-age hypothesis. They must say that God created stellar clusters in such a way that they appear as if they had experienced an evolutionary history with a calculable duration—not an authentic history in

---

16. Steidl, *The Earth, the Stars, and the Bible* (Grand Rapids: Baker, 1979), p. 153.

which events really occurred but an illusory history that never actually took place. In the end, advocates of the young universe theory must accept a universe that is much like a library of fantasy, filled with books that give intricately detailed accounts of events that never happened. While the apparent-age concept does eliminate the need to wrestle with several difficult questions concerning the theological implications of the evolutionary character of cosmic history, it also transforms the Creator of a coherent cosmos into a divine magician—a master of the deceptive art of illusion.

If the cosmos were not Creation, perhaps an illusory cosmic history would be acceptable. But cosmic history is the history of the Creation—the unfolding of the intentions and purposes of the Creator. Therefore, if cosmic history is illusory, the intentions and purposes of the Creator must be illusory as well. The Creator revealed by Scripture, however, is no mere magician, and his purposeful intentions are no mere illusion. We have every right and responsibility to expect that the Creation—the handiwork of the Creator—will provide evidence for an authentic history only. We can assume that the Creation will display both integrity and coherence in all of its properties, all of its behavior, and in all of its history. To expect less of the Creation is to expect less from the Creator.

Its Inadequate Concept of Creation

The apparent-age hypothesis is not derived from any explicit biblical teaching; it is rather an inference drawn from an inadequate concept of creation—the notion that creation is identical to instantaneous inception by divine fiat.

The roots of the idea that creation is limited to instantaneous inception must be identified with a particular approach to the interpretation of Genesis 1 and other scriptural references to God's creative activity. The proponents of special creationism have chosen to treat Genesis 1 as if it were a journalistic account of a sequence of specific past events rather than to recognize it as an artistic illustration of an eternal covenantal relationship extending over the past, the present, and the future. This approach fails to recognize the covenantal function of Genesis 1 in the scriptural canon and thereby fails to see its function as the preamble to the covenantal canon that introduces the sovereign as God the Creator and establishes the status of the whole cosmos as his Creation. The proponents of special creationism have failed to distinguish the vehicle from the packaging and content of scriptural literature, and so have been led to direct a host of inappropriate questions concerning the material world to Genesis 1 and the rest of the Bible. They have demanded that Scripture supply answers to questions of cosmic history and cosmic chronology—questions it was never meant to answer.

The product of these misperceptions is the assertion that the essence of creation (as a divine act) is nothing more than instantaneous inception by divine fiat. God's work as Creator has been isolated from all of cosmic history, including the present moment, and confined to an initial act of exnihilation. It would seem, then, that the essence of knowing God as Creator is merely to know that he performed some specific feats in the past.

Those special creationists who wish to have the "creation model" taught in the public school science classroom have formulated an even more restricted and theologically barren concept: *creation,* as the term is used in that context, means merely "instantaneous inception"; even the reference to divine fiat has been stripped away.

When the heart of the biblical doctrine of creation—the faithful and purposeful covenantal activity of God the sovereign Creator—has been removed, what is left is no more than a corpse. What its proponents call "scientific special creationism" is a theologically empty concept that does not deserve to be associated with the religiously rich doctrine of creation. The biblical concept of creation as a dynamic covenantal relationship to the faithful Creator ought never to be replaced by the notion of a magical act of instantaneous inception performed by an unspecified magician.

Its Violation of the Integrity of Natural Science

"Scientific creationism" represents an attempt to employ natural science in such a way that it appears to provide warrants for belief in the concepts of special creationism.[17] But since it has reduced the concept of creation to the notion of a recent in-

17. I place the terms "scientific creationism," "creation science," and "scientific creationist" in quotation marks because they represent concepts I judge to be markedly different from what the words themselves might suggest. As our discussion will illustrate, I judge these concepts to promote neither legitimate natural science nor a theologically acceptable definition of creation.

stantaneous inception of a mature, functioning universe by an unspecified agent, "scientific creationism" has little else to do but to argue for a young universe and seek to discredit the concept of evolutionary processes, which it considers to be inherently naturalistic. Consequently, its argumentation is almost entirely negative in character (recall Morris's assertion that "evidence against evolution constitutes positive evidence for creation"). Thomas G. Barnes, whose theory of terrestrial magnetism we will take a look at shortly, claims that "it takes but one proof of a young age for the moon or the earth to completely refute the doctrine of evolution"[18] and that "there is nothing more devastating to the doctrine of evolution than the scientific evidence of a young earth age."[19] Thus, one of the principal arguments of "creation science" is this: If the cosmos is young, then it didn't have sufficient time to evolve; therefore it must have been created.

Is there any physical evidence to support the hypothesis that the earth (the entire planet, not merely some sample of it) is no more than few thousand years old? "Scientific creationists" have been engaged in a vigorous search for such evidence. But they have set themselves up for falling into an all-too-familiar trap, one that we have all slipped into at one time or another. People who "know" the answer to some question and then search for evidence to support that predetermined answer will be strongly tempted to see only supportive evidence, while overlooking, or even choosing to disregard, any evidence to the contrary. In theology that might be called "hunting for prooftexts"; in natural science it would soon be identified as biased investigation, or just plain poor science.

An interesting peculiarity of the apparent-age hypothesis is that while age can presumably be merely a superficial appearance, youth must be real. The logic is quite simple. Authentic history is presumed to contribute to the appearance of age in the ordinary manner. Thus the smallest value for age derived by empirical methods must be closest to the actual duration of cosmic history. Consequently, the method that yields the smallest value for apparent age (preferably in the neighborhood of ten thousand years) will be given a special status.

---

18. Barnes, "Young Age for the Moon and Earth," Impact no. 110, Institute for Creation Research, August, 1982.

19. Barnes, "Earth's Magnetic Age: The Achilles Heel of Evolution," Impact no. 122, Institute for Creation Research, August, 1983.

A favorite among the arguments for a young earth is Barnes's theory concerning the earth's magnetic field and its temporal development.[20] Measurements of the strength of the dipole component of the terrestrial magnetic field over the last 150 years reveal that it has been steadily diminishing at the rate of approximately five percent per century during that period. In what appears to be an intense desire to "prove" that the earth is only a few thousand years old, Barnes has advanced a hypothesis concerning the cause and history of the magnetic dipole field. He postulates that it is being caused by electrical currents circulating in the conductive core of the earth. Furthermore, he postulates that the earth's electromagnet was energized at "the creation" and has been steadily running down ever since. He assumes that there is no driving mechanism in operation today, so the electrical energy is being steadily dissipated by ordinary resistive heating.

If this model were accurate, then the strength of the terrestrial magnetic dipole field would decay in an exponential manner. Therefore, Barnes has sought to fit the observed field decay to an exponential function. Doing this, he has determined the half-life of the dipole field to be about 1,400 years: every 1,400 years the earth's field will be cut in half. Thus, for every 1,400 years that one goes back in history, the earth's magnetic field, claims Barnes, must be doubled. Going back more than ten thousand years yields, he says, "inconceivably large" values for the terrestrial field— too large for human comfort. Thus, concludes Barnes, "one can show that the earth's magnetic age is limited to thousands of years, not the billions claimed by evolutionary scientists."[21]

To a person who is not trained in the physical sciences, Barnes's theory may sound like a reasonable approach. It is, however, a travesty of scientific investigation, violating the integrity of natural science. Let us look briefly at the major flaws in Barnes's methodology.[22]

---

20. In addition to the articles cited in the two previous notes, see Barnes's "Depletion of the Earth's Magnetic Field," Impact no. 100, Institute for Creation Research, as well as ICR Technical Monograph No. 4, "Origin and Destiny of the Earth's Magnetic Field," 1973.

21. Barnes, "Earth's Magnetic Age," p. 1.

22. For a more detailed discussion and critique of Barnes's theory, see chapter 8 of Davis A. Young's *Christianity and the Age of the Earth* (Grand Rapids: Zondervan, 1982), and Stephen G. Brush's essay "Ghosts from the Nineteenth Century: Creationist Arguments for a Young Earth," in *Scientists*

First, his assumption that the present dipole field is the remnant of a decaying primordial magnetic field is unwarranted. He chooses to assume that there are no processes now occurring that could generate the earth's magnetic field, so that what we now observe must be merely the decaying remnant of the original, divinely imposed electromagnet. There are, however, several driving mechanisms presently under investigation that might account for the phenomenon, and the fact that geophysicists are honest enough to admit no universal agreement on which combination of these best accounts for all of the observed properties and behavior of the terrestrial field does not warrant Barnes's cavalier dismissal of the entire concept. Personal bias does not constitute a sufficient warrant for such a wholesale dismissal of several reasonable models for magnetic field generation.

Second, Barnes's fitting of an exponential function to the empirical data is entirely inconclusive. A linear function fits equally well,[23] but since it would yield a "magnetic age" of millions of years and fails to meet the assumptions concerning a lack of driving mechanism, he gives it no consideration. Actually the time period covered by the data that Barnes employs is far too brief to warrant extrapolation beyond a century or two. His conclusions, therefore, being based on unlimited extrapolation, are not only unjustified but entirely meaningless.

Third, and most disappointing, Barnes has chosen to ignore an entire body of knowledge about the history of the terrestrial magnetic field. When rock is formed by solidification from the molten state or by a sedimentation process, information about the intensity and direction of the magnetic field present at the time of formation may be permanently locked into the material. Paleomagnetic studies of this natural remanent magnetism reveal that the terrestrial magnetic field has not experienced the history postulated by Barnes. Instead, the physical record clearly shows that the earth's magnetic field has undergone cyclic variations in both intensity and direction during the past several hundred million years. Even on a much shorter time scale,

---

Confront Creationism, ed. Laurie R. Godfrey (New York: W. W. Norton, 1983). For critiques of other "scientific creationist" methods and arguments, see *Science and Creationism,* ed. Ashley Montagu (New York: Oxford University Press, 1984), and Philip Kitcher's *Abusing Science* (Cambridge, Mass.: MIT Press, 1982).

23. See Brush, "Ghosts from the Nineteenth Century," p. 74, and Young, *Christianity and the Age of the Earth,* p. 119.

archeological studies reveal that the field intensity reached a peak about two thousand years ago and had only one-half its present value about six thousand years ago.[24]

Barnes's decision to base his model on an unwarranted assumption regarding the absence of any driving mechanism, his dependence on an inconclusive extrapolation of selected empirical data, and his conscious decision to dismiss an entire body of physical data that directly contradicts his hypothesis vividly demonstrates what happens when you begin an "investigation" already "knowing" the outcome and then proceed to search for evidence to support the preconceived conclusion. Such a methodology does not represent the excellence in scholarship in the natural sciences toward which Christians should strive.

I could cite many other examples of unacceptable methodology being practiced by "scientific creationists." The astronomic investigation of cosmic history is dismissed by most such individuals as a meaningless study of the "superficial appearance of history." Radiometric dating of terrestrial, lunar, and meteoritic samples—one of the most reliable means of pinpointing the time at which certain events in the history of the solar system history and the earth occurred—is dismissed in the same cavalier manner as Barnes's dismissal of paleomagnetic studies. The second law of thermodynamics is employed to give the impression that evolutionary processes are physically impossible, but when the incompleteness and fallacious character of their argumentation is pointed out by competent scientists, "scientific creationists" turn a deaf ear.[25] To those who understand thermodynamics as it is applied to cosmic history, it is abundantly clear that the second law neither forbids nor ensures that a process such as biological evolution will occur.

The common thread in nearly all examples of "scientific creationist" argumentation is a highly selective approach to the employment of natural science. When an argument based upon some empirical data or some standard scientific methodology favors the desired (and preconceived) result, that argument is then employed with unrivaled strictness and vigor. If, however, there are theoretical considerations or empirical data that spoil the favored argument, then they are either ignored or summarily dismissed. "Scientific creationism" takes a cafeteria-style ap-

---

24. See Young, *Christianity and the Age of the Earth*, p. 123.

25. See John W. Patterson, "Thermodynamics and Evolution," in *Scientists Confront Creationism*, pp. 99-116.

proach to natural science: take only what you like and pass by the rest. Legitimate science, however, is never permitted to use this selective approach. All data must be incorporated, and all theoretical principles must be given their due consideration. Christian scholarship is not exempt from these demands; on the contrary, it is called upon to set the example of honesty and openmindedness in the performance of excellent natural science.

"Scientific creationism" has the superficial appearance of a science, but in reality it is a syncretistic assembly of unwarranted assumptions, inconclusive argumentation, capricious rejection of unfavorable theoretical principles, and the cavalier selection of favorable data—all dedicated to the verification of preconceived conclusions. It gives little respect to the integrity of natural science—and does not do much better concerning the scriptural message it ostensibly honors. It gives the appearance of promoting belief in the biblical doctrine of creation, but in reality its concept of creation has been stripped of its biblical content concerning the covenant relationship between the Creator and his Creation, leaving only a meaningless statement about the instantaneous inception of the cosmos by an undesignated agent.

I am fully convinced that "scientific creationism" is a travesty of natural science and a sad parody of biblical theology. It is not my intent to condemn the leadership of the movement, but I cannot in good conscience fail to reject its concepts and methodology. It appears that persons motivated by a sincere desire to praise God as their Creator have become trapped in an unbiblical approach to the realization of that desire. I heartily commend the desire but wish to suggest a more biblically faithful manner for expressing the highest praise of our Creator-Redeemer.

CONCLUSIONS

Let's close this chapter by underscoring three key points.

1. The concepts of "creation" and "evolution" constitute answers to entirely different questions.

"Creation" and "evolution" are not contradictory answers to the same question; each speaks to distinctly different issues. "Creation" is the answer to the question of what status the material world has. "Evolution" is an answer to the question of what material process characterizes the temporal development of the universe. "Evolution" provides answers to a set of questions that belong in the category of "internal affairs," while "creation" provides answers to the basic questions in the category of "external relationships." The concepts of creation and evolution are not contradictory. Nor do they provide supplementary answers to a common set of questions. Rather, they provide appropriately different answers to categorically complementary questions.

2. The contemporary creation/evolution debate is a tragic blunder.

The either/or format of the contemporary debate is entirely fallacious; it promotes a false dilemma spawned by the fallacy of many questions. Naturalistic evolutionism and special creationism are not the only two possible positions regarding the status and history of the cosmos.

The current debate propagates a misunderstanding of the concepts of both creation and evolution. "Creation" does not necessarily entail the instantaneous inception of a fully developed universe. The Genesis narratives and other biblical materials on creation may better be seen as artistic illustrations of the eternal covenant relationship of Creator and Creation than as journalistic reports of specific past events. And "evolution" is not an inherently naturalistic concept. According to biblical theism, all material processes are divinely governed and directed; evolutionary processes are no exception.

3. The authentic debate is the wholly religious antithesis between atheistic naturalism and biblical theism.

"Scientific special creationists" are seeking to convince the public that a legitimate debate exists between two "sciences"—"creation science" and "evolution science." Such a contrivance, however, is the unfortunate consequence of promoting the illegitimate invasion of the domain of natural science by a literalistic strain of biblical exegesis. In actuality, the creation/evolution debate cannot legitimately be conducted in the scientific laboratory or the science classroom.

"Biblical special creationists" are seeking to present the creation/evolution debate as a legitimate confrontation of science and religion. But that effort is the product of a misunderstanding of both creation and evolution, and is the consequence of allowing both natural science and biblical exegesis to invade the exclusive territory of the other.

There is an important and authentic debate to be carried out, but it lies entirely in the religious domain. While "creation" and "evolution" are not themselves competing answers to the same question, the two worldviews with which the contemporary debate (incorrectly) associates these concepts are indeed diametrically opposed. The worldviews of atheistic naturalism and biblical theism represent antithetical religious commitments. The authentic antithesis is a purely religious confrontation between theism and atheism. It is most unfortunate that this authentic confrontation has become obscured by the fog of the creation/evolution debate. The real debate is not between creation (perceived merely as instantaneous inception) and evolution (a natural process of continuous development), but between biblical Christian faith and modern Western naturalism. That debate ought to be carried out with full vigor and clarity, unencumbered by the distraction and obfuscation that has been introduced by the shouting match between the proponents of instantaneous inception and the defenders of continuous development.

CHAPTER TWELVE

The Creationomic Perspective

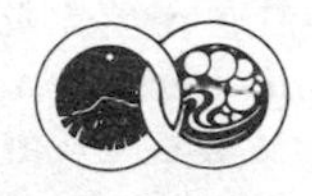

A THIRD OPTION

In our evaluation of the creation/evolution debate, we unhesitatingly rejected the position of naturalistic evolutionism—not because of any principial objection to the concept of evolutionary natural processes but because of the naturalistic world-view into which it places all natural processes, including evolution. We reject naturalism because it fails to see the cosmos as Creation, because it fails to look at the cosmos through the spectacles of Scripture.

We also rejected the position of special creationism. While it does promote a concept of creation, it fails to present the full biblical concept. In place of the dynamic, convenantal relationship for all time it substitutes a discrete set of singular events in the past. Similarly, while it does proclaim the cosmos to be Creation, it presents a created world the properties, behavior, and history of which are quite different from those that we actually observe; in place of coherence, continuity, and authenticity it substitutes incoherence, discontinuity, and a montage of illusory appearances.

And in addition to rejecting both of the principal positions in the contemporary creation/evolution debate, we also rejected the either/or format of the confrontation itself, dismissing the entire debate as a tragic blunder. Having made all of those assertions, we are obligated to offer a third option—an option that suffers neither the fatal flaws of naturalistic evolutionism nor the pitfalls of special creationism.

But is there such an option? Indeed there is. Drawing on terminology I introduced earlier, I'm going to call it "the creationomic perspective."

The creationomic perspective is achieved when natural science is placed in the framework of biblical theism. The foundation of this perspective on the cosmos and its history is the recognition that the entire cosmos is God's Creation. Its chief methodological rules are the principles of categorical complementarity. The creationomic perspective calls for questions concerning the relationship of the Creation to its Creator to be directed to the Bible, and it demands that the Bible's answers be taken seriously. Those who adopt the creationomic perspective perceive all natural phenomena as lawfully governed by the Creator. And having the creationomic perspective will lead us to take the Creation seriously when careful scientific investigation yields the answers to a host of questions concerning its specific properties, its coherent patterns of behavior, and its authentic physical history.

THE CREATIONOMIC PERSPECTIVE ON THE COSMOS

Its Standing Relative to God

When Genesis 1 (or any other biblical reference to creation) is read in the awareness of its function in the covenantal canon and in the light of a familiarity with its cultural and historical context, its central message becomes evident: God is the sovereign Creator of all things, and the cosmos stands not as another deity, not as a competitor to God, not as a substitute for God, but as Creation—the contingent servant of its Creator. As the Creator, God acts as the Originator, Preserver, Governor, and Provider of his Creation. The essence of the Bible's revelation concerning God as Creator is that God has established with his Creation an everlasting covenant—an eternal, dynamic relationship of faithful action. God is the eternal, faithful, sovereign Creator; the cosmos, through all of cosmic history, is his Creation.

The Implications of the Status of the Cosmos

We can draw a number of conclusions from the biblical revelation concerning the Creator-Creation relationship. Because God is its Originator, the Creation has a radically contingent existence—an existence that is totally dependent on God's creative power. And because God is its Preserver, or Sustainer, the Creation manifests a continuity of existence (its existence is continuously sustained rather than repeatedly re-created). Furthermore, because the Creation is a product of divine activity, the material world possesses an authentic reality—it is no mere illusion of the human mind or senses.

Many expectations concerning the general character of material behavior follow from the biblical teaching that the cosmos is not autonomous but governed and directed by the Creator. We expect the behavior of matter and material systems to be orderly and patterned. We expect the behavior of the cosmos to exhibit coherence and integrity. We expect the behavior of the universe to be discernible and intelligible to the human mind. We expect the behavior and the properties of matter to be correlated so that material behavior will exhibit proximate cause-effect relationships. And because cosmic history is divinely directed, we expect the cumulative effect of material phenomena to display evidence of purposeful, directed development. Because the Bible teaches that God is the Governor of all phenomena in his Creation, we can assume that all natural processes are manifestations of the Creator's faithful and purposeful governance; no natural phenomenon whatsoever falls outside of the domain of divine governance.

Cosmic Status and Scientific Investigation

Because natural science is the disciplined investigation of the properties, behavior, and history of the material world, certain basic aspects of scientific methodology and attitude will follow from our expectations concerning the character of Creation. As natural scientists, we will be encouraged to search for order in both the structure and behavior of matter and material systems. We will be encouraged to promote theoretical models that incorporate the idea of coherence and integrity within the system of natural laws that describe material behavior. We will not expect to find the sort of irrational or capricious behavior that a sentient, independent cosmos might exhibit. We will be encouraged to postulate a continuity of causally related events and processes and to search for proximate cause-effect relationships. And we will accept cosmic history as authentic and purposeful. We will be predisposed to reject any suggestion that either the cosmos or its history is a mere illusion. In essence, because the cosmos has the status and character of Creation, we are encouraged to take it seriously in the manner that natural science does.

In *The Clockwork Image,* Donald MacKay makes a very similar point:

> The scientific approach . . . is not just uneasily compatible with biblical theism, but rather positively encouraged by it. The events of our physical world are, from the Christian standpoint, the continual gifts of a faithful and non-capricious Creator. . . . The mechanistic thought-models of modern science thus spell out in one particular form the very same trust in the faithfulness of God that the Bible inculcates more generally in the man of faith.[1]

THE CREATIONOMIC PERSPECTIVE ON COSMIC HISTORY

Evolution Defined

Like so many other terms that we have used, the term *evolution* is used in various contexts to convey a considerable diversity of meanings. To prevent misunderstandings in the discussion that follows, let's review our working definition of the term.

> *Evolution is the change in the properties (e.g., physical, spatial, chemical, biological) of a material system (e.g., a planet, star, galaxy, universe) or of a sequence of material systems (e.g., the successive generations of living organisms) that occurs as a consequence of natural processes that take place entirely within the system or that transpire in response to an interaction between the system and its environment.*

In light of that definition, I think a few comments are in order.

1. Evolution is an ordinary natural process—a process that is not fundamentally different in character or status from other natural processes, such as a summer sunrise, a winter snowstorm, the blooming of a flower, or the birth of a child.

2. No natural process is inherently naturalistic.

3. To accept the concept of evolutionary processes does not require the introduction of phenomena that go beyond the ordi-

---

1. MacKay, *The Clockwork Image* (Downers Grove, Ill.: InterVarsity Press, 1974), p. 88.

nary pattern of material behavior. Rather, the concept of evolution represents an extension of our present experience of continuity into the indefinite past (and future?). Furthermore, the concept of evolution removes the arbitrary imposition of discontinuity and incoherence that is demanded by the notion of instantaneous inception.

4. To describe cosmic history as evolutionary in character is to say no more than that the behavior and temporal development of the universe is orderly, coherent, causally continuous, and purposefully directed—the same qualities of behavior that we expect as a consequence of the character of divine governance.

Appreciating the Evolutionary Character of Cosmic History

In the creationomic perspective, evolutionary processes have no special status relative to other divinely governed natural processes, and they shouldn't be accorded any special treatment. Nevertheless, they have been so commonly misrepresented as inherently naturalistic that the Christian community has had little practice at employing an appreciation for these phenomena in our praise of the Creator. With David, for example, we sing "The heavens declare the glory of God, the vault of heaven proclaims his handiwork" (Ps. 19:1), but I suspect that few of us have sung "The expanding universe declares the glory of God, and cosmic evolution proclaims his handiwork." The sentiment, however, is precisely the same. Employing concepts and vocabulary appropriate to our respective cultures, both we and the psalmist are praising God as the sovereign Creator by whose design all things function as they do and by whose purposeful direction all things have come to be as they are. As David was responsible for giving praise to God for all the divine works of which he had knowledge, so we are called to return praise to our Creator-Redeemer for all of his activities, even for that which we have discovered by the scientific investigation of his Creation—even for cosmic evolution.

What follows is by no means an exhaustive list of the theological implications of cosmic evolution. I simply want to touch on some of the positive consequences of appreciating the evolutionary character of cosmic history; I would challenge others, particularly theologians in the evangelical, and Reformed segments of the Christian community, to extend these efforts far beyond what I am able to provide here. Help us all to incorporate the discoveries of the twentieth century into our understanding of where we stand in the history of God's work as the Creator and Redeemer of the cosmos and into our songs of praise for all that he is doing.

1. Dynamic Order and Divine Creativity

During the last few centuries, the concept of the order displayed by the cosmos has changed from one of static order to one of dynamic order.[2] Many ancient and medieval thinkers conceived of cosmic order in terms of static structures, fixity, changelessness. Some even thought of change itself as evidence of imperfection. But in recent centuries, particularly the past two, inquiry into the nature of the Creation has revealed that the old static concept of order was incomplete at best. While many material systems or species of creatures do appear to possess stable characteristics over relatively short periods of time (the lifetime of a human observer, for example), careful investigation has revealed that nearly all material systems, even species of living organisms, do undergo change over extended periods of time. Stars, for instance, long thought to be the epitome of constancy and changelessness, are now known to undergo processes of change. The evidence convincingly illustrates that stars evolve through a sequence of stages from gravitationally collapsing protostars to gravitationally crushed remnants such as white dwarfs, neutron stars, and black holes. There is still a great deal of order in this drama, but it is the *dynamic* order of coherent patterns of development, not the *static* order of fixity or changelessness.

Cosmic history is like a magnificent tapestry woven from different strands of temporal development to form the intricately designed pattern of cosmic evolution. Spatial evolution, galactic evolution, elemental evolution, stellar evolution, planetary evolution, and biological evolution are coherently integrated and intertwining processes that serve as the individual threads in the tapestry: the dynamic order of patterned development marks the whole of cosmic history. That such a vast spectrum of structures and creatures—from galaxies to galagos, from star clusters to starfish, from habitable planets to human beings—could be formed according to a single set of coherently related patterns for material behavior simply dwarfs the human

---

2. See N. Max Wildiers, *The Theologian and His Universe* (New York: Seabury Press, 1982).

imagination and the capacities of human creativity. Surely the dynamic order of patterned behavior exhibited by the Creation throughout its expansive history serves as an inexhaustible tribute to the boundless vastness of divine creativity. Would it not require far greater creativity to design and direct the dynamic processes that constitute cosmic evolution than simply to mandate the existence of the end product alone?

2. The Integrity of Creation and the Unity of God

The Bible teaches that God is a unity. In God there is no duplicity, no internal inconsistency or contradiction, no lack of integrity. The Creation offers countless examples of this divine unity and integrity. Like its Creator, the Creation displays the coherence of an Ansel Adams landscape photograph rather than the random juxtaposition of unrelated elements as in some bizarre collage. The cosmos displays an integral unity of substance and behavior: it is made of one kind of matter, it behaves according to one set of rules, and its patterned behavior is systematically correlated with its material properties.

The evolutionary pattern of cosmic history unites all material systems and all natural processes or events. Our bodies, for instance, are quite literally made of "the dust of the earth." But the earth and the other planets of our solar system were formed as by-products of the birth of the sun out of the collapsing solar nebula. Terrestrial planets, rich in the heavy elements, are made principally from materials that were manufactured billions of years ago in massive stars that lived and died explosively long before the sun was born. Hydrogen and helium, the raw materials that fueled the production of these heavy elements, are themselves the material remnant of the singular burst of activity that marked the genesis of the visible universe. Furthermore, the properties of the spacetime framework in which we now live and move on planet earth are established by the cosmic distribution of all material bodies—even of those galaxies and quasars located at distances too great for the human mind to comprehend.

The scientific investigation of cosmic history has revealed a most remarkable quality of the Creation: the existence of all celestial bodies in the cosmos and the occurrence of all events in cosmic history actively contribute to what exists and happens here and now. Or, to state it even more strongly, if the Creation that we now observe is real rather than illusory, and if the cosmic history that is written in the observable properties of the

Creation is authentic rather than superficial, then the entire corporeal cosmos and the whole of its evolutionary history are essential for our existence at this moment. From the evidence gathered, we conclude that the whole of cosmic history is far more than just a fascinating story; it is a *requirement* for the material reality of our flesh and bones.

What is the source of the integrity, the coherence, and the unity displayed by the cosmos and its evolutionary history? Naturalism must look to the cosmos itself for the source of these qualities. Biblical theism, which provides the framework for the creationomic perspective, looks to the Scriptures. I must confess that I do not understand what satisfaction a naturalist can draw from metaphysical speculation on the origin of the cosmos and its qualities, but I do know the comfort and assurance that can be derived from the Bible's revelation that the entire cosmos and all of its qualities have their origin in God the Creator. The integrity, coherence, and unity displayed by the Creation is a manifestation of the unity of the Creator himself.

3. Natural Law and Divine Faithfulness

In the creationomic perspective, natural laws are held to be statements describing the patterned behavior that matter and material systems exhibit as a consequence of divine governance. Natural laws are not prescriptive laws *of* Nature for its own behavior but descriptive representations of the laws of God *for* nature, which is his Creation. Therefore, we may expect the character of natural law to reveal something of the character of the law-giver, the Governor of Creation.

Recall some of the statements that we have made concerning the character of material behavior and the natural laws describing that behavior. We noted the substantial body of evidence supporting the assertion that natural laws are spatially and temporally invariant. We noted that natural processes exhibit a causal continuity, that the state of the cosmos at any given moment is causally connected to the states immediately preceding and following—thereby connecting past, present, and future. We noted that the laws of natural behavior are coherently and consistently related to one another. And we have seen that cosmic history is not an arbitrary assemblage of independent events but rather a patterned succession of related phenomena. The evolutionary pattern of cosmic history displays integrity and coherence in every conceivable way.

Do these qualities of natural law, material behavior, and

cosmic history come as any surprise to one who sees the cosmos as Creation? Indeed not. The Creator of whom the Bible speaks is the God who has entered into a covenant agreement with his Creation. He has promised to be faithful—never arbitrary, never capricious. Natural laws, which describe the Creator's patterns of governance, manifest with utmost clarity the covenantally ensured faithfulness of the Creator. By its coherence, cosmic history illustrates that God does not act arbitrarily. And by the continuity of its patterns of evolutionary change over all ponderable time, cosmic history shows that God does not act capriciously.

4. An Evolving Creation as the Context for Human Responsibility

From cover to cover, the Bible calls humanity to obedience; human beings, unique among all creatures, are capable of choosing between right and wrong, and are held accountable for their choices and actions. But would the concept of human responsibility be equally legitimate and meaningful in any sort of material cosmos? Is there something special about a cosmos with an evolutionary history in this regard? The answer is not immediately obvious. The answers to significant questions seldom are.

Before attempting to address this difficult question, let's pause to remind ourselves of some of the characteristic properties of this cosmos—the one whose history is evolutionary in character. In discussions to this point we have concluded that our cosmos is characterized by objective reality, lawful governance, coherence, unity and integrity, continuity, purposefully directed development, and authentic history. Obviously, this is not an exhaustive list, but it does identify those properties most relevant to this particular discussion.

Now what about this matter of human accountability? In what kind of a world is human responsibility a legitimate and meaningful concept? In a world in which objective reality is replaced by superficial appearance? In a world in which material behavior is not lawfully governed? In a world characterized by incoherence or internal contradiction? In a world that lacks unity or integrity? In a world whose history is punctuated by discontinuities? In a world that manifests no evidence of purposefully directed development? In a world whose history is illusory rather than authentic? I think not. I judge that a world fitting any of these descriptions would lack a quality that is essential for providing a meaningful context for human responsibility. But note that a world in which cosmic history is evolutionary in character need not lack any of these qualities. A Creation characterized by objective reality, lawful governance, coherence, unity, integrity, continuity, purposeful development, and authentic history provides a fitting context for both evolutionary development and human responsibility.

There remain those who argue that the concepts of evolutionary development and human responsibility are contradictory or mutually exclusive. Viewed from the creationomic perspective, however, the concepts are not contradictory but singularly compatible. To consider the possibility that we are creatures (members of God's Creation) whose capacity for the awareness of self, of God, and of our responsibility for obedience of divine mandates has been formed through a process of continuous evolutionary development does not strike me as inappropriate or incongruous or unbiblical. I see no reason whatsoever to deny that the Creation might have an evolutionary history or that morally responsible creatures might have been formed through the processes of evolutionary development.

EDUCATIONAL IMPLICATIONS: TEACHING ABOUT THE CONCEPT OF EVOLUTION

Silence Is No Solution

In the climate created by the resurgent creation/evolution debate, the question of how, or whether, to present the concept of evolution (usually restricted to a consideration of its biological aspect only) has become a highly volatile issue in educational circles. Approaches spanning a broad spectrum are being proposed, some with the expenditure of much emotional energy. Public and private schools alike struggle to find an approach that is educationally sound and at the same time satisfactory to their constituencies. It is not an easy task for school boards, administrators, or teachers. And the difficulty of making acceptable decisions is compounded by the either/or confrontational approach that has characterized the creation/evolution debate.

One way to avoid favoring any particular manner of presenting the concept of evolution is simply to eliminate the topic from the curriculum. If there exists a diversity of strongly held opinions on the way evolution (or any other topic, for that matter) should be presented, then adopting any particular approach is almost certain to dissatisfy some segment of a school's constituency. Therefore, some people argue, why not just eliminate the problem by eliminating the topic altogether?

This approach poses an insurmountable problem of its own, however: it is, quite simply, educationally irresponsible. One of the principal tasks of educators is to equip students with both the knowledge and the skills to make intelligent judgments on a whole range of issues. It may be that one of the reasons the adult community is now having such a difficult time dealing with the subject of evolution in the educational curriculum is that they themselves never received an adequate education concerning the scientific and philosophical issues involved. If that is the case, surely it would be educationally irresponsible to ensure the perpetuation of ignorance by choosing to avoid the issue by leaving it out of the next generation's curriculum. To teach a general astronomy course without treating the topic of cosmic evolution or a general biology course without treating the topic of biological evolution would be educationally indefensible; it would be like teaching literature without reference to poetry or American history without reference to the Revolutionary War or mathematics without mention of the process of multiplication. The silence would create a monumental gap in knowledge. Avoiding the topic of evolution may be an easy way to avoid the unpleasantness of facing tough decisions, but it would be cowardly and self-defeating in the long run.

If the issue cannot be avoided, how might it be responsibly treated? Let's evaluate some of the approaches that are vying for adoption.

Ways of Presenting Evolution in Public Schools

1. As a Naturalistic Process?

Using the assumptions of naturalistic science. We have already defined "naturalistic science" as natural science placed in the framework of a thoroughly naturalistic worldview. Naturalistic science explicitly makes the claim that *all* natural processes are inherently naturalistic: whatever happens is the consequence solely of self-existent matter behaving according to self-governed patterns. Evolution, like any other material process, is viewed as an autonomous process that proceeds without any need for divine governance. According to naturalism, the material world is all there is: the cosmos is Nature—a machine-like substitute for God. Naturalistic science is reductionistic: human beings are *nothing but* marvelous molecular machines, and evolution is *nothing but* a mindless, mechanical manifestation of Nature's undirected, purposeless self-expression. Naturalistic evolutionism is the consequence of elevating the concept of evolution from a descriptive model for a natural process to an explanatory framework for a naturalistic worldview.

Clearly, the public educational system ought not to teach evolution in the framework of a commitment to naturalism. The public schools should do their utmost to maintain a neutrality with respect to religious commitment. Therefore, naturalism, which constitutes a very definite (atheistic) religious commitment, ought not to be propagated. The public educational system ought not to favor one religious commitment over another either overtly or in subtle or subversive ways. The obvious conclusion, then, is that *naturalistic evolution should not be taught in the public school system.*

Using the "two-science" model. One of the aims of the Institute for Creation Research and many other "scientific creationist" organizations is to have the public school system present evolution only in the context of what they call the "two-science" model. The two "sciences" are labeled "evolution science" and "creation science." Various descriptions of "creation science" make it abundantly clear that it is intended to be a presentation of scientific evidence for the instantaneous inception of the universe in the manner envisioned by special creationists. "Evolution science," on the other hand, is simply naturalistic evolutionism.

The heart of the whole "two-science" approach is the either/or format of the creation/evolution debate. The proponents of this approach insist that there are only two "models" of origins—creation (defined as instantaneous inception of a fully developed cosmos) and evolution (defined as naturalistic evolutionism)—and since the public school system may not favor either of the two, then, in the interest of fairness, it should present them both. But this appeal to fairness doesn't hold up. As we have already seen, naturalistic evolutionism and special creationism are not the only two ways of accounting for the status and temporal development of the universe (nor, indeed, is either even accurate). The claim that only these two particular perspectives need be presented is simply fallacious—for all of the same reasons that the creation/evolution debate itself is a fallacious and fruitless exercise.

My objections to the "two-science" model, however, are based not only on the fallacy of its either/or format, but also on the fact that the two-model approach grants to naturalism its entirely unwarranted claim that evolutionary processes are inherently naturalistic. I do not wish to have my children, or any other children, taught that the processes of evolutionary development necessarily exclude the concept of divine governance and that these phenomena must be associated exclusively with an atheistic perspective on natural processes. No Christian should tolerate the propagation of such a naturalistic perspective. Whether as the sole position or as an alternative to special creationism, *naturalistic evolution should not be taught in the public school system.*

2. As a Natural Process?

According to the principles of categorical complementarity, the investigation of natural processes lies fully within the domain of the natural sciences, as does the study of natural history, which is simply the cumulative effect of such processes.

By our definition, a natural process is any material process (whether divinely governed or self-governed we do not say) that occurs in conformity to the ordinary patterns for material behavior. In the spirit of that definition, evolutionary processes have no special status; they are no more (or less) extraordinary than those that provide us with a magnificent sunset over Lake Michigan, a tulip blooming in May, or the metamorphosis of a creepy caterpillar into the winged flower of a butterfly.

Consequently, the study of evolutionary processes and the evolutionary character of cosmic history is both a legitimate activity for the natural scientist and a legitimate topic for discussion in science courses such as astronomy or biology—and not only legitimate but mandatory, because the investigation of cosmic evolution as it is manifested in the temporal development of galaxies, stars, planets, and all other celestial systems is such an important component of contemporary astronomy that it would be irresponsible to exclude it from a general astronomy course. Similarly, because the study of biological evolution is such an important aspect of contemporary biology, excluding it from a general biology course would amount to an act of intellectual and educational irresponsibility.

The topic of evolution, however, like any other legitimate topic in the natural sciences, must be taught in a manner that honors the boundaries of the scientific domain. Only those aspects of evolution that deal with questions concerning the properties, behavior, and temporal development of material systems fall within the scope of natural science. As long as discussion is confined to these questions of "internal affairs," the topic of evolutionary development can be treated in a religiously neutral manner—a manner that is mandatory for the public school science classroom.

Questions of governance and other matters that involve the relationship of the material world to external, nonmaterial powers or beings should be treated separately. Such questions lie outside the domain of natural science and therefore should be discussed in a different setting—one in which matters of worldview and religious perspective are the focus of attention. Though one religious perspective ought not to be favored over another in the public school classroom, it would be entirely appropriate to formulate fundamental questions about the status, origin, governance, value, and purpose of the material world in such a context. Students deserve to be shown the distinction between questions of internal affairs and external relationships. They deserve to have it explained to them that only questions dealing with internal affairs should be treated in their natural science courses. Students deserve to be exposed to a clear presentation of the *questions* concerning the status of the cosmos and the consequences of that status, and they should be encouraged to seek answers to these questions from the source of their own worldview or religious perspective.

Thus, the topic of evolution ought not to be ignored in the public school curriculum; rather, it should be presented in a way that respects the diversity of religious perspectives in its constituency. Naturalistic evolution ought not to be taught either as the sole perspective or as the only alternative to special creationism—which is to say that evolution should be discussed in a way that honors the principles of categorical complementarity. Questions regarding evolution as a natural process may be discussed in natural science courses, but students should be encouraged to take questions concerning the status, origin, governance, value, and purpose of the universe and its evolutionary history to the sources of their own worldview or religious perspectives. I am fully convinced that this approach is educationally sound and that it clarifies the question of religious perspective in a way that does not favor any one particular worldview.

Ways of Presenting Evolution in Christian Schools

1. Evolution as an Alternative to Creation?

One of my greatest fears is that a majority of Christian children are being taught to view creation and evolution as alternatives. I am afraid that the either/or format of the contemporary creation/evolution debate provides ample evidence that millions of Christian people, including many educators, continue to accept the claims that evolution is an inherently naturalistic process and that creation necessarily entails the instantaneous inception of a fully developed universe, complete with the superficial appearance of history. And by presenting naturalistic evolutionism and special creationism as the only two alternatives, the two-science model leads children to associate the results of mainstream natural science with naturalism and to link the Christian faith with an account of cosmic history that is contradicted by an abundance of empirical data.

If we proceed to raise our children in the "school of either/or," teaching them in our Christian schools that they must choose between creation and evolution, we burden them with a vexing dilemma. When some of them go on to receive further education in the natural sciences, when they become more knowledgeable concerning both the methods and results of scientific investigation and gain an appreciation for the validity of empirical data and the credibility of its interpretation in terms of a coherent, evolutionary cosmic history, what kind of choice can they make?

Children have long been raised in this fashion, and so we need not wonder what the results of such an education might be. Some individuals thus taught continue to cling to special creationism, apparently willing to live with a persistent disparity between what they see and what they believe. Others, willing to embark on a difficult journey, choose to reject the basic assertions of the either/or approach and seek an approach that will allow them to take both the Bible and the Creation seriously. While I admire the tenacity of belief that the first choice requires, I admire even more the courage of faith and the persistence of effort required by the second. (And, optimistic dreamer that I am, I expect that such individuals will ultimately arrive at the principles of categorical complementarity, perhaps an improved version of what I have struggled to formulate.)

There is, however, a third option—a tragic one taken all too often: not a few individuals reject the Christian faith and embrace the tenets of a naturalistic worldview. The line of reasoning is straightforward. (1) According to the either/or approach, one must choose to believe *either* in evolution (which implies the truth of naturalism) *or* in special creation (which is a required component of Christian faith). (2) Empirical evidence overwhelmingly supports the conclusion that the cosmos has experienced an evolutionary history spanning billions of years. (3) Therefore one is obligated by intellectual honesty to choose naturalistic evolutionism. This tragically false conclusion is the unfortunate consequence of beginning with the false premise that evolution must be presented to our children as an alternative to creation. This is a tragedy that we can—and must—avoid.

2. Evolution as a Method Employed by God?

In their attempts to understand the place of evolution in the Christian perspective, many Christians have referred to evolution as the method that God has employed in the formation of his Creation and his creatures. While I strongly prefer this viewpoint to that of the either/or approach, I do have some reservations about the appropriateness of referring to evolution, or to any other material process, as a "method employed" by God to accomplish his purposes.

Perhaps I am merely quibbling over semantics, but it seems to me that to speak of natural processes such as thermonuclear fusion and photosynthesis and biological evolution as methods employed by God suggests that these phenomena exist as independent, autonomous entities and that God somehow manages to coax them into his service. The "method employed" language lends itself to the interpretation that the natural process is the primary reality and that God's employment of it is secondary and contingent upon its prior autonomous existence. It seems to me much preferable to view God's action as the primary entity and to see natural processes as the expression of that reality. This puts all natural processes in their proper place—under divine governance. It is for this reason that I have tried consistently to refer to natural processes not as "methods employed" by God but rather as the "consequences of divine action" or the "manifestations of divine governance." Similarly, I prefer to speak of cosmic evolution not as the "method employed" by God to form his Creation but as the "expression of God's strategy" for the temporal development of his Creation.

In connection with this, I would like to stress that I am not advocating what some have called "theistic evolution." In fact, the term *theistic evolution* strikes me as a prime example of the "methods employed" language. The natural process is given primary status by its appearance in the form of the noun *evolution;* divine action appears to be relegated to a secondary status by its confinement to the adjective *theistic.* "Theistic evolutionism," therefore, is not an appropriate label for the position that I am advocating. I prefer to stick with "the creationomic perspective." Divine action ought always to be the matter of primary importance, with natural process as the secondary or contingent reality.

This emphasis on the primacy of divine action should not, however, be interpreted as implying that God deterministically causes every event that occurs in his Creation. This Creation is no mere puppet, and its Creator is not some sort of divine puppeteer whose control eliminates the possibility for responsible freedom of action on the part of his creatures. Within the dynamic order of the divinely governed Creation and its divinely directed history there appears to be a remarkable balance between predictable regularities and meaningful contingencies. Natural history is permeated with events that are contingent on circumstance and with occasions in which choice must be exercised by responsible creatures. As we noted earlier, the Creation of which we are a part is a fitting context for human responsibility.

3. Evolution as the Ongoing Expression of God's Strategy for Creation

There is not the slightest doubt in my mind that the observed patterns of material behavior are a manifestation of God's faithful governance and that the course followed by cosmic history is the consequence of God's providential direction. Neither is there any doubt in my mind that natural science is doing an intellectually honest and excellent job of describing both the patterns of material behavior and the course of cosmic history. The logical conclusion that I must reach, then, is that the evolutionary character of cosmic history that scientific investigation has uncovered represents the continuous expression of God's strategy for the temporal development of his Creation.

How, then, shall we present the concept of evolution in the Christian school classroom? Everything I have said thus far points to the fact that evolution and all other natural processes occurring in the material world should be studied in the framework of creationomic science. Viewing the universe and its temporal development from the creationomic perspective, we perceive the whole cosmos as the Creation: its existence originates in God, its behavior is governed by God, its value lies in its covenantal relationship to God, and its history expresses the purposes of God. From the creationomic perspective we see evolutionary processes neither as alternatives to divine creation nor as preexistent powers employed by the Creator but as manifestations of divine governance and ongoing expressions of divine providence. Evolutionary processes are ordinary natural processes having no special status relative to other material phenomena.

In the classroom of a Christian school we can feel free to deal with all categories of questions about the material world. Questions pertaining to its internal affairs can be addressed to the natural sciences, and the results of scientific investigation can be welcomed and respectfully evaluated by the customary standards of intellectual inquiry. Natural science should be treated as a friendly ally of the Christian faith, not as an enemy to be feared, scorned, or crushed. Furthermore, questions that pertain to the relationship of the material world to God its Creator can be addressed to the Bible, which is the proper source of answers to such questions. In the Christian school classroom, both the Bible and the Creation can be taken seriously. Appropriate questions can be addressed to each and the answers they provide can be respectfully incorporated into a fully theistic worldview.

EDUCATIONAL IMPLICATIONS: TEACHING ABOUT THE CONCEPT OF CREATION

Ways of Presenting Creation in Public Schools

The concept of creation, like that of evolution, has become a topic of considerable concern in the educational arena. One common line of argumentation claims that if evolution is taught, then creation should also be presented. I have suggested that the concept of the evolutionary history of the universe is too important to omit and that while its presentation might raise some difficult problems, silence is indefensible. What about creation, then? Must the concept of creation also be presented in the public schools? I would say Yes—but only on the condition that it be presented appropriately.

My reason for adopting this position has nothing to do with whether or not the concept of evolution is taught. I am not making an appeal for "equal treatment" for the simple reason that "creation" and "evolution" are quite separate issues—they represent answers to entirely different questions. Each must be considered on its own merits. My reason for proposing that the concept of creation be incorporated into the public school curriculum is that I consider it to be a culturally and historically important issue that all educated people should be familiar with, regardless of whether it is a part of their particular worldview.

But as I said, I support this point of view only on the condition that creation be presented in an appropriate manner. Unfortunately, the most prominent current proposals for introducing creation into public school curricula would have it presented in ways that are wholly inappropriate. Let's take a look at the two principal approaches.

1. Creation as an Outmoded Religious Concept

Modern Western naturalism, the most prevalent nontheistic worldview in our American culture, tends to think of creation as an ancient religious (or mythological) concept that has been rendered obsolete by the discoveries of modern natural science. This attitude has, unfortunately, been reinforced by the persistent claims made by special creationists that creation necessarily entails the recent, instantaneous inception of a fully developed universe that merely appears to be old. But I believe that this perspective on the concept of creation betrays a misunderstanding of both natural science and creation.

Presenting the concept of creation as an ancient *religious* belief that has been replaced by a modern *scientific* perspective is entirely unacceptable: no authentic religious concept, ancient or modern, can be replaced by a scientific concept, ancient or modern. Substitution is impossible because science and religion deal with essentially different categories of questions, and neither can provide sensible answers to questions that ought appropriately to be addressed to the other.[3]

3. Some may argue that, contrary to my claim, substitutions of this kind are quite commonplace in the history of human thought. A. D. White's *History of*

2. Creation as a Scientific Concept

The "two-science" model for origins that has been proposed as a means of giving equal time to (allegedly) alternative concepts treats the topic of creation as if it were a scientific concept. "Creation science" engages in a search for empirical warrants for belief in the instantaneous inception of a mature universe.

I completely reject this approach for reasons I have already given: first, because creation ought never to be treated as an alternative to evolution, and second, because creation is intrinsically a religious rather than a scientific concept. To attempt to deal with the concept of creation in a nontheistic manner seems to me as absurd a project as trying to account for the formation of the Grand Canyon without reference to the action of the Colorado River.

3. A Third Option: Creation as a Meaningful Religious Concept

Creation, we have noted, is chiefly a concept concerning the relationship of the material world to God. The concept of creation provides a very specific kind of answer to the questions of status, origin, governance, value, and purpose. Creation is a thoroughly theistic concept and it must be presented as such—or not at all.

In a public school setting, no one religious perspective may be favored above others. Therefore, the extent to which the concept of creation may be discussed is quite limited. I would hope, however, that it would be possible (perhaps in a unit on comparative religion or cultural perspectives or the like) to introduce the whole set of categories of questions on the external

---

the Warfare of Science with Theology in Christendom (1896), for instance, is filled with illustrations of religious belief being replaced by scientific description or explanation. But such critics should take a closer look at the nature of the belief being replaced. I suggest that in nearly every case White cites, what was being replaced was a belief concerning material phenomena that was mistakenly thought to be warranted by biblical exegesis. These were beliefs held for religious reasons, but they were not authentic religious beliefs about religious matters. The categories have been confused for centuries!

If there is a lesson to be learned, it is that biblically warranted religious belief must provide answers to authentically religious questions (e.g., questions concerning relationship to God) and must not be allowed to invade the legitimate domain of natural science. As we noted earlier, when we address inappropriate questions to Scripture, we will get nonsensical answers in return. However, when we address appropriate religious questions to the Bible, natural science is powerless to challenge its answers.

relationships of the material world and to present the concept of the Creator-Creation relationship as one meaningful perspective. I would hope that it would be possible to describe the fundamental differences between theistic and naturalistic worldviews (both being thoroughly religious in character) and to point out how the concept of creation contributes to a theistic perspective. And I would hope, moreover, that creation could be presented as a concept not entailing a particular chronology or a particular scenario for cosmic history but rather as a source of answers to the questions regarding cosmic status and related issues.

Am I hoping for too much? Perhaps so. But if in the context of public education we cannot provide the answers, we ought at least to clarify the questions and to encourage our students to seek the answers from appropriate sources. The distinction between questions of internal affairs and questions of external relationship is foundational, and it is essential that we understand it. Our students deserve to be shown which questions may legitimately be addressed to the scientific investigation of the cosmos and which questions fall outside of the domain of natural science. Our students deserve to be guided in formulating genuinely religious questions and to be encouraged to take these questions to the sources of their own worldview or religious perspectives. Indoctrination in particular answers is forbidden, but guidance in clarifying questions and encouragement in seeking answers is mandatory.

Presenting the Biblical Doctrine of Creation in the Christian School Classroom

Perhaps this part of our discussion on the educational implications of the creationomic perspective is superfluous. In the Christian school classroom it is both our privilege and our responsibility to teach all subjects in the framework provided by a biblically theistic worldview. Therefore, any study of the material world, whether in a natural science course or in some other setting, will be carried out in the awareness that the object of study is God's Creation.

I wrote this entire book for the purpose of clarifying what it means to take the Bible seriously when it speaks about the cosmos as God's Creation, what it means to take the Creation seriously when we engage in a scientific investigation of its nature, and how to synthesize the views of the cosmos as seen through the spectacles of Scripture and the lens of natural science. In a real sense, then, everything that has been said contributes in some way to my answer to the question of how we should present the concept of creation to our children—at home, in our church educational programs, and in the Christian school classroom. In closing, then, I would simply like to underscore a few key points pertaining to the biblical doctrine of creation.

1. As the covenantal canon, having the status of Word of God, the Bible's principal function is to bring us into a proper covenantal relationship with God, with one another, and with our material environment. Not only functionally, but also structurally, the Bible is a thoroughly covenantal document.

2. The biblical doctrine of creation provides authoritative answers to questions concerning the status, origin, governance, value, and purpose of the material world—all matters involving the relationship of the cosmos, humanity included, to God.

Declaring that the cosmos stands under God as his Creation, the Bible boldly contradicts both ancient naturalistic polytheism and modern materialistic naturalism; the cosmos, says the Bible, is neither an assembly of deities (or their visible manifestations) nor an independent autonomous substitute for deity.

3. The biblical doctrine of creation does not outline any specific model of natural history or cosmic chronology.

The Bible employs a variety of literary genres and figures of speech when speaking of God's activity as the Creator. These statements ought not to be reduced to technical descriptions or primitive scientific specifications concerning the mechanism of divine origination, preservation, governance, or providence. We should recognize that Genesis 1, for example, is not a journalistic chronicle of past events but an artistic illustration of an eternal relationship. In its role as the opening statement of the covenantal canon, Genesis 1 serves to introduce the covenant God as the Creator and to establish the status of the entire cosmos as his Creation. Bringing questions about physical properties, material behavior, natural history, or cosmic chronology to the Bible is simply inappropriate to its very character and its purposes.

4. The study of the physical properties, material behavior, and temporal development of the cosmos is both exclusively and exhaustively the domain of the natural sciences.

I say "exclusively" because the Bible does not deal with questions in these categories—questions concerning the internal affairs of the universe without reference to their origin in divine action. I say "exhaustively" because the natural sciences can do no more. Questions concerning the relationship of the material world to nonmaterial powers or beings lie outside of the domain of natural science.

5. The views of the cosmos we get through the spectacles of scriptural exegesis and the lens of natural science are categorically complementary. The two views provide answers to categorically complementary sets of questions about the material world, God's Creation.

6. The complementarity of these two views has implications for the way in which the concept of creation is taught.

First, *the concept of creation cannot be presented as the result of scientific investigation.* Creation is a thoroughly theistic concept and therefore lies outside of the domain of natural science. "Creation science" represents neither competent natural science nor sound biblical exegesis.

Second, *the biblical doctrine of creation ought not to be presented as an alternative to the results of natural science.* Because scriptural exegesis and natural science deal with mutually exclusive sets of questions, the answers to one set cannot sensibly be offered as an alternative to the others. The biblical doctrine of creation, for example, is not an alternative to the scientific theory of evolution. Presenting creation and evolution as if they were mutually exclusive answers to the same question is entirely fallacious and constitutes an unbiblical either/or-manship.

Third, *the biblical doctrine of creation should be presented as the framework in which excellent natural science can be performed to the glory of the Creator.* This is the heart of the creationomic perspective. It keeps natural science in its proper place—and what better place is there for natural science than the place of its birth, where it may continue to be nurtured and guided by the biblical doctrine of creation. It is this doctrine that embodies the biblical revelation that God is the sovereign Creator and that the entire cosmos is his Creation; that as Creator, God is our Originator, Preserver, Governor, and Provider; and that as Creation, the cosmos can be expected to exhibit coherent, causally continuous behavior and authentic, purposefully directed history. In such a Creation it is possible for us to engage in honest scientific investigation. Far more importantly, in such a Creation we are called to live our whole lives in the awareness that we are responsible creatures of the covenantally faithful Creator-Redeemer.

Recommendations for Further Reading

ON GENESIS AND CREATION

Because the topics of creation and the book of Genesis are so important, I find it difficult to select only a few books for recommendation. Thus, the works I cite here must be viewed as forming only the beginning of an adequate list of worthwhile books.

Genesis must, of course, be read in the context of the entire Old Testament. As a general reference work capable of illuminating that larger context, I suggest Bernhard W. Anderson's *Understanding the Old Testament,* 3d ed. (Englewood Cliffs, N.J.: Prentice-Hall, 1975). Nahum Sarna's *Understanding Genesis* (New York: Schocken Books, 1970) provides us with a Jewish perspective on the first book of the Bible. For discussions focused on the early chapters alone, see Henri Blocher's *In the Beginning,* trans. David G. Preston (Downers Grove, Ill.: Inter-Varsity Press, 1984); Claus Westermann's *Creation,* trans. John J. Scullion (Philadelphia: Fortress Press, 1974); and Alan Richardson's *Genesis 1-11* (London: SCM Press, 1953). Meredith Kline's *The Structure of Biblical Authority,* 2d ed. (1972; rpt., Grand Rapids: Eerdmans, 1975) provides valuable insights into the canonical character of the Bible, and Leland Ryken's *The Literature of the Bible* (Grand Rapids: Zondervan, 1974) serves as an excellent introduction to the wealth of Scripture's literary qualities. For a helpful discussion on the relationship of Genesis to modern science, I recommend the first four chapters of Conrad Hyers's *The Meaning of Creation* (Atlanta: John Knox Press,

1984). A lucid and thorough study of the Christian doctrine of creation will be found in Langdon Gilkey's *Maker of Heaven and Earth* (Garden City, N.Y.: Doubleday, 1959).

ON ASTRONOMY AND COSMOLOGY

There are many good astronomy textbooks on today's market that will provide the motivated reader with a substantial introduction to contemporary astronomy. Representative of these are Jay M. Pasachoff's *A Brief View of Astronomy* (New York: CBS College Publishing, 1986), Michael A. Seeds's *Foundations of Astronomy* (Belmont, Cal.: Wadsworth, 1984), and William J. Kaufmann III's *Universe* (New York: W. H. Freeman, 1985). For an introductory text that includes more astrophysical detail and deals candidly with the unsolved puzzles in contemporary astronomy, see Frank H. Shu's *The Physical Universe* (Mill Valley, Cal.: University Science Books, 1982).

Those wishing to go beyond the brief discussions of physical cosmology found in general textbooks such as those cited above are advised to look at Edward R. Harrison's *Cosmology: The Science of the Universe* (Cambridge: Cambridge University Press, 1981). Though this book was written for the nonspecialist, I have found it to be one of the most informative and well-written books on the topic. Harrison's latest book, *Masks of the Universe* (New York: Macmillan, 1985), provides the general reader with an insightful and provocative perspective on the history of the human search for knowledge of the universe and the meaning of existence. For a Christian perspective on the interaction of theology and cosmology from the Middle Ages to the present day, see N. Max Wildiers's *The Theologian and His Universe,* trans. Paul Dunphy (New York: Seabury Press, 1982).

Timothy Ferris's *The Red Limit,* 2d ed. (New York: Quill, 1983) is an extraordinarily well-written history of the development of 20th-century cosmological concepts. *The First Three Minutes* (New York: Basic Books, 1977), written by Steven Weinberg, is the landmark publication that first unveiled to the general public the "standard model" for the early history of the big-bang universe.

ON THE SCIENCE-RELIGION RELATIONSHIP

The vast area encompassed by the topic of the relationship between science and religion is probably best entered by looking at some of its history. A good starting point is *Religion and the Rise of Modern Science* (Grand Rapids: Eerdmans, 1972), by R. Hooykaas. Standing in the same tradition is Colin A. Russell's recently published *Cross-Currents: Interactions between Science and Faith* (Grand Rapids: Eerdmans, 1985). Persons willing to tackle a more ponderous work should consult Stanley L. Jaki's *The Road of Science and the Ways to God* (Chicago: University of Chicago Press, 1978).

The remainder of this list is composed of books on the science-religion relationship written by Christian authors during the last three decades. The list is by no means exhaustive; my apologies to those many authors whose worthy contributions do not appear. I shall proceed alphabetically by author.

Ian Barbour's *Issues in Science and Religion* (New York: Harper & Row, 1966) provides one of the best overviews of the historical, methodological, and religious issues at stake in the larger discussion; his *Myths, Models, and Paradigms* (New York: Harper & Row, 1974) is a stimulating discussion of the role of models in both natural science and religion. Richard H. Bube's *The Human Quest* (Waco: Word Books, 1971) employs the concept of levels of understanding to demonstrate that scientific descriptions and religious descriptions of any phenomenon do not compete with one another. David L. Dye, in his *Faith and the Physical World* (Grand Rapids: Eerdmans, 1966) takes the position that physical data are philosophically neutral, so that the Christian need never choose between the Scriptures and scientific data. Persons seeking a brief and nontechnical discussion that aims to reduce the perceived tension between biblical studies and natural science may wish to look at Robert B. Fischer's *God Did It, but How?* (Grand Rapids: Zondervan, 1981).

At a rather different level, Malcolm A. Jeeves's *The Scientific Enterprise and Christian Faith* (Downers Grove, Ill.: InterVarsity Press, 1969) is based on papers presented at a 1965 conference on science and Christian belief held in Oxford, England. The stated purpose of Jeeves's book is "to explain why science is a true friend of biblical faith." Well known and admired for his clear and insightful writing on the topic of science and faith, Donald M. MacKay has written three brief books that are well worth the reading: *Science and the Quest for Meaning* (Grand Rapids: Eerdmans, 1982), *The Clockwork Image* (Downers Grove, Ill.: InterVarsity Press, 1974), and *Science, Chance, and Providence* (Oxford: Oxford University Press,

1978). For a more extensive development of the creation-science-faith relationship, see Eric Mascall's *Christian Theology and Natural Science* (New York: Longmans, Green, 1956) or A. R. Peacocke's *Creation and the World of Science* (Oxford: Clarendon Press, 1979), representing the Bamptom Lectures of 1956 and 1978.

For thirty years Bernard Ramm's *The Christian View of Science and Scripture* (Grand Rapids: Eerdmans, 1956) has provided evangelical Christianity with sound advice. Though its science is now three decades out of date, and though I am not comfortable with Ramm's model for progressive creation, I maintain a high regard for the role that this book has played in encouraging careful thought on the topic at hand. Another insightful book of about the same vintage is Aldert van der Ziel's *The Natural Sciences and the Christian Message* (Minneapolis: T. S. Denison, 1960).

Let me end this list with two recent publications. First, *The Genesis Connection* (Nashville: Thomas Nelson, 1983), by John L. Wiester. Although I am uncomfortable with some of the questions that Wiester addresses to Genesis 1, I think he provides a good overview of terrestrial history, and he encourages the Christian community to appreciate its grandeur. Finally, I would highly recommend Davis A. Young's *Christianity and the Age of the Earth* (Grand Rapids: Zondervan, 1982) to Christians who remain uncomfortable with the multibillion-year chronology of cosmic history.

CRITIQUES OF "CREATION-SCIENCE"

For further discussion and evaluation of the "creation-science" movement, I would recommend beginning with Roland Mushat Frye's *Is God a Creationist?* (New York: Scribner's, 1983). Although I'm not entirely happy with the title, I consider this to be an excellent assembly of essays, all written by people who profess their faith in God as Creator, which collectively provide a powerful and thoughtful critique of the "creation-science" approach to natural science and biblical interpretation. Philip Kitcher's *Abusing Science* (Cambridge, Mass.: MIT Press, 1982) provides a thorough analysis from the philosophical perspective. Other valuable essays, written from a diversity of perspectives, can be found in *Scientists Confront Creationism* (New York: W. W. Norton, 1983), edited by Laurie R. Godfrey, and *Science and Creationism* (New York: Oxford University Press, 1984), edited by Ashley Montagu.

Ronald L. Numbers has written a concise history of the development of "creation-science," published under the title "Creationism in 20th-Century America" in the 5 November 1982 issue of *Science,* pp. 538–44. For a forthright exchange of perspectives between proponents of "creation-science" and scientifically trained Christians who hold views similar to mine, see *Creation and Evolution* (Leicester: InterVarsity Press, 1985), edited by Derek Burke. This volume is part of a series of debates in print entitled When Christians Disagree, edited by Oliver R. Barclay.

Index of Names and Subjects

Index of Scripture References

Old Testament

New Testament